YOUNG SCIENTIST USA.

Vol.4

Copyright © 2015 by Authors

Layout by Gilbert Rafanan, Paul Bourianow
Cover by Alina Panchenko

All rights reserved. In accordance with the U.S. Copyright Act of 1976, the scanning, uploading, and electronic sharing of any part of this book without the permission of the publisher constitute unlawful piracy and theft of the author's intellectual property. If you would like to use material from the book (other than review purposes), prior written permission must be obtained by contacting the publisher.

The publisher is not responsible for websites (or their content) that are not owned by the publisher.

3702 W Valley HWY N
 STE 204-31245
Auburn, WA
98001

http://www.YoungScientistUSA.com/

Printed in the United States of America

Lulu, 2015

ISBN 978-1-329-72471-6

Table of Contents

Social Science

Economics and Business

Residential Mortgage Development in Russia With Regard to International Practices. 3
Olga Zhukova

How to Be a Good Manager: Basic Behavioral Competencies. 10
Yana Kochkova

«Resource Curse»: the Subject of Discussion or Inevitability of Search for Future Energy Resources?. . . 14
Kambar Musabekov, Shavkat Sunakbayev

State Support for Small Businesses in the Chechen Republic. 18
Muslim L. Tamaev

Education

The Role of National Upbringing in Formation of National Ideology of the Youth. 23
Umida Boyzokova

Methods of Ecological Knowledge Training of Pupils and Celebration of the Festival "Day Of Birds" within the Framework of the Out-of-Class Activity. 25
Sulgun Durdyeva

Role of Labour Education in the Ethnopedagogics in the modern world. 29
Nailya Zaydullina

Mathematical modeling of production processes using MS EXCEL. 33
Elena Mirzoyeva

Quality Management of Education in Child Welfare Centers. 36
Anatoly Red'ko

The Issues of Formation of the Graphic Literacy among the Teachers of Fine Art
and Engineering Graphics. 41
Erkin Ruziev, Rustam Latipov

Law

Administrative Legal Relationships and Administrative Regulations in the Russian Legislation
as Components of Administrative Treatment Mechanism Concerning Illegal Use of Drugs. 47
Hui Wang

Citizenship in International Law: Concept and Legal Content. 51
Ksenia Tyurenkova, Karine Muravieva

Other Social Sciences

Comparative Analysis of Parameters of the Functional Readiness at Students
of the Meliorative College. 57
Galina Khasanova, Regina Niyazova

Humanities
Arts

Glimpses of History of the Formation of the Orchestra of Turkmen National Musical Instruments. . . . 63
Bahar Goshayeva

Applied Science
Engineering

Autonomous System for Monitoring the Integrity of Navigation Data Provided by Satellite Navigation Systems Based on Optimal Information Processing Algorithms for Navigation Systems of Land Moving Objects. 71
Alexander V. Ivanov, Dmitry Komrakov, Dmitry Boykov

Thermal Insulation Materials in Designing Communications-Electronics Equipment for Hypervelocity Vehicles. 79
Yevgeny Rodikov

Control of Network Standby Power Supply Enabled Once the Islanding System Opens the Sectionalizing Circuit-Breaker. 83
Leonid Surov, Ilya Phillipov, Igor Fomin

Medicine

The Choice of Pathogenetic Treatment of Chronic Generalized Periodontitis. 89
Andrey Sushchenko, Olga Oleynik, Elena Vusataya, Oksana Krasnikova, Elena Alferova

Current Aspects of Parkinson's Disease Treatment. 95
Vyacheslav Kutashov, Natalia Kameneva, Maxim Yurin, Nikolay Zhuchkov

Current Peculiarities of Diagnosis and Treatment of Benign Paroxysmal Positional Vertigo. 98
Vyacheslav Kutashov, Valentina Matviets

Conversion in Laparoscopic Cholecystectomy. 102
Margarita Ryzhikova, Anna Soloviyova

Osteoplastic amputation in the middle third of the thigh. 105
Ryzhikova Margarita, Soloviyova Anna

Post-stroke Depression. 108
Alexander Shulga, Vyacheslav Kutashov, Marina Shulga

Agriculture

Innovative Technology of Harvesting of Cotton Stalks. 113
Merdan Shammedov

Other

Academic Background. 119
[1]Anzor Amadaev, Dzhabrail Dasaev, Artur Amadaev

Forecast of potential natural risks for the historical and architectural sights of the Holy Mountains Lavra (Sviatohirsk Monastery) 126
Valeriy Sukhov

SOCIAL SCIENCE

Economics and Business

Residential Mortgage Development in Russia With Regard to International Practices

Olga Zhukova

Russian State University of Physical Education, Sport, Youth and Tourism (SCOLIPE), Moscow

Abstract: The innovative system of residential mortgage management (ISRMM) suggested by the author is based on the compact coexistence of a non-bank deposit-credit organization (NDCO) operating under the Russian Central Bank's license and construction business represented by builders and developers within a holding structure. Introducing the ISRMM in Russia based on European construction companies' practices will promote the idea of a social mortgage bank and facilitate developing and implementing innovative mechanisms of social mortgage.

Keywords: mortgage, innovation, social, AHML, commercialization, economics.

Innovative systems of residential mortgage management based on European construction companies' practices [1]

The evolution of innovative mortgage mechanisms in Russia goes hand in hand with the development of national management structures and population-oriented credit and deposit services market. In studying this issue the author focuses especially on non-bank credit organizations.

Mortgage risk management in non-bank credit organizations as a subsystem of national finance deserves scrutiny, for it determines the choice of the very system of managing the institutional environment.

The innovative system of residential mortgage management (ISRMM) suggested by the author is based on the compact coexistence of a non-bank deposit-credit organization (NDCO) operating under the Russian Central Bank's license and construction business represented by builders and developers within a holding structure. Such a management system is called a homocluster.

Functionally this structure bears resemblance to construction companies in Great Britain. The basic lending services provided by constructing companies include, first and foremost, consumer lending (mortgage and consumer loans), managing the savings of individuals and legal entities, issue of credit cards, insurance services, retail operations.

The effectiveness of deposit and credit servicing and cash management of a constructing company results from the effectiveness of its financial mechanism, which predetermines the forms and the methods of transforming the population's savings into investments.

One of constructing companies' major competitive advantages is their ability to accumulate the needed initial contribution and provide financial resources to the company participants in form of loans as well as to finance house-building for their participants.

It was the German system of mortgage lending that the author took as an example of construction savings-banks, which basically are construction companies as well, although they are categorized as specialized banks. In Germany the source of funds for residential loans in construction savings-banks consists of citizens' deposits, borrowers' payments for taking the loan, and state premium granted in accordance with the contract on construction savings provided that a depositor has topped up his or her deposit during the year. A mortgage loan may be granted exclusively to a construction savings-bank depositor on the expiry of a certain period (usually a few years) during which he or she has been monthly depositing a sum stipulated in the contract. The interest rates (both for deposits and loans) are fixed for the whole duration of the contract. Generally they tend to be about 4% lower than the market ones.

Currently commercial banks are exposed to serious competition from European construction companies both in mortgage lending and in deposit services. The majority of constructing companies in Britain aim their activities above all at managing money flows and supporting liquidity of credit institutions, while acquiring financial results comes in the second place. Their services may be called social by nature of their aims since construction companies do not seek profits.

Today there are no loan consumers' cooperatives (LCC) in Russia, although the Agency for Housing Mortgage Lending (AHML) used to regard them as primary lenders. The author maintains that their place should be taken by non-bank credit organizations as specialized banks for mortgage services in the housing market.

The infrastructure of non-bank credit relations in European developed lending markets is represented by an array of specialized financial and credit institutions which provide financial support of economic agents' activities by delivering traditional specialized services. It also serves as an institutional environment for the development of national financial services markets. Real estate market is one of the most important segments of institutional environment. At the same time, it cannot function properly unless mortgage market is stable.

Many aspects of the identified research problem have been covered in works of economists from such a perspective that they substantiate initiatives to improve the non-bank credit organizations' financial mechanism. However, they do not take the merging of real estate market and mortgage lending market into consideration. These markets have common consumers and investors and, therefore, the author posits that their structure of management must fully meet the modern requirements made by financial practice.

An objective necessity may turn to the objective reality only as a result of comprehensive study and improvement of non-bank credit institutions' financial mechanism and infrastructure, and increased operation efficiency due to diversification of financial, credit and investment services, particularly in mortgage lending market.

Resolving the issue of forming the resource base for financial and credit activities of credit organizations in Russia has facilitated the trend towards more common approach of the construction and financial branches to financial and credit mortgage lending operations that take into account the experience of British mortgage market.

The present period of non-bank construction companies development in Britain is grounded on mortgage lending to clients with average levels of income who buy houses and apartments within low and average price range.

The conservative financial and economic policies of non-bank construction companies (long-term mortgage loans, punitive sanctions for early redemption, etc.) allow them, unlike American mortgage lending sphere, to retain the needed level of financial stability amid the global financial crisis. The 2008-2009 crisis hit the US mortgage lending market. Russia has always been committed to developing mortgage market according to American standards promoting a two-level system.

Today the extensive ways of construction companies' development in Great Britain have exhausted themselves. The coming modernization will focus on cutting the operational expenditures, reducing the administrative costs and reinvigorating the activities in financial time markets. As a result, these companies' activities will be more universal and diversified.

Taking into account the advantages and the drawbacks of British construction companies and German construction savings-banks the author suggests that Russia should develop the innovative system of residential mortgage management (IS-RMM) with a non-bank deposit-credit organization (NDCO) as the holding's core company.

The Russian Central Bank has in its turn taken into account the facilitating and counteracting factors of organizing investment and operating money flows and prudently separated NDCOs by function (deposit and credit activities without operating functions) from NCOs (operating functions without deposit and credit activities). Thus, the risks of mixed activities inherent in banks have been reduced.

NDCOs have been set to develop specialized financial services, and it is these services that the author suggests employing to the full in order to develop innovative systems of residential mortgage [2].

It is worth noting that there are no mortgage NDCOs in Russia as they offer the same mortgage products as banks in the housing market and cannot compete with them in amounts of current assets.

At the same time, NDCOs have more abilities to attract investments than housing savings-cooperatives because they have more guarantees and better access to cheap interbank loans, although the author still considers it insufficient for winning a client in competitive struggle.

Currently legal entities are registered in the AHML as agents and attract clients by lower interest rates on mortgage loans as their risks are smaller than those of banks. Their income consists of commission fees from clients (for completion of a loan and risk insurance) and from the AHML (as a compensation for supplying mortgages). The share of such agent companies in the mortgage market is insignificant, so banks do not view them as rivals.

The author supports the reasonable assumption that the infrastructural aspect of non-bank credit institutions development expands the field for lending through involving more participants.

Applying innovative systems of mortgage management results in cheaper loans and better access to lending sources, which brings about an increase in financial effectiveness due to the growing inflow of borrowers. Based on this the ISRMM may be justly categorized as a social mortgage bank. Its aimed functioning in the mortgage and real estate market will provide productive turnover of real economy resources, which can contribute to effective credit policy in Russia.

Introducing the ISRMM in Russia based on European construction companies' practices will promote the idea of a social mortgage bank and facilitate developing and implementing innovative mechanisms of social mortgage. This factor will change the situation in the Russian lending market.

A social mortgage bank as a model of the innovative mortgage management system in mortgage market opens up prospects for realizing the opportunities mentioned above. Today the Russian Central Bank and the State Duma's bills are increasingly promoting their practical implementation.

Having formulated the main operating principles of the socially-oriented ISRMM the author points out the commercial appeal of these ideas to small and medium businesses in banking, mortgage lending, housing construction and insurance.

The institutional development of residential mortgage in view of international integration processes [3]

Approbation of ideas and exchange of experience should be attributed to international integration processes in research and innovation. For instance, development of innovative mortgage mechanisms always implies references to similar practices in Europe, USA and Canada. Sprouts of new phenomena always emerge on well-fertilized grounds of ideas. Hence the enduring achievements of interstate and interuniversity exchange of research results and implementation practices.

Residential mortgage is an open complex institutional system susceptible to the influence of the environment and the endogenous processes. Studying the institutional essence and peculiarities of the present-day residential mortgage in Russia and revealing its internal problems and dysfunctions increases scientific validity of its reformation both at the macroeconomic and local levels.

The introduction of the American pattern of residential mortgage development implies developing the market of mortgage-backed securities and active involvement of international actors in the mortgage capital market.

The mortgage crisis in the USA (2007-2009) actuated various pressure groups related to the mortgage business in Russia and formed negative expectations both among creditors and borrowers, which even in 2015 prevents lending rates from falling. These mortgage market trends are rather significant since the risks of negative expectations are included in the loan costs.

Vagueness of the institution environment in this sphere influences the effectiveness of Russia's mortgage market as well. There are no necessary amendments to the law "On Mortgage-Backed Securities", construction savings-banks are developed insufficiently, and the mortgage sphere is developing without regard to the European experiences.

Forming the funding base is crucial for residential mortgage. Therefore, the enhancement of tools and institutions for attracting long-term financial resources is gaining importance in terms of mortgage market development.

Mortgage-backed securities constitute a major source of residential mortgage funding. It is attracting additional resources from securities markets that has brought down the value of the securities in developed countries. By the same token, this has led to reduced cost of mortgage loans.

As a result, mortgage-backed securities are turning into the main mechanism for attracting capital into housing building.

Developing mortgage and lending mechanisms is an important priority for Russia, and it will contribute to resolving social, economic and financial problems. According to social studies, nearly 77% of the Russian population would like to improve their housing conditions. Proceeding from the fact that 10% have enough money at their disposal to buy housing with a mortgage loan, total estimated capacity of the mortgage market presently comes to at least 5 mil. mortgage loans. Taking into account that an average loan (including the regions with smaller loans) amounts to $20 000, the potential size of the market has already reached at least $100 billion.

Mortgage lending in Russia cannot develop properly without mortgage-based securities market. Securitization of mortgage presents a real opportunity to reduce mortgage lending risks for financial institutions. Introduction of new tools in form of mortgage-backed securities into the Russian financial market will help to solve the problems of market condition. It is an open secret that Russia lacks tools for investment in conditions of high ruble liquidity and its strengthening exchange rate.

Studying the foreign experience is of significant interest for Russian practice. The securities market is now in the making. Mortgage mechanisms and their ways of introduction are being perfected, and the legal framework is being adjusted to the new market mechanisms. It will result in establishing conditions for the proper operation of these mechanisms. However, the issue of securities has been sporadic.

One has to admit that the institutional imbalance in the branch has a negative effect on the Russian economic development. Substantial analysis of transplantation and adaptation of mortgage institutions in emerging economies can be found only in the works of V. Polterovich, O. Starkov and E. Chernyh. However, these researchers tackle only specific aspects of analysis leaving many other issues related to the institutional substance of mortgage behind (such as institutional disparities in price formation, etc.)

Summing up the problems of mortgage development in emerging economies, the author may point out the major ones:
- low solvency of the population and a large share of shadow economy without stated income;
- lack of long-term cheap resources in the local capital market;
- inefficient judicial procedures related to foreclosure and sale of mortgage;
- low motivation of banks to develop mortgage lending caused by the need for much effort;
- inadequate development of the market's infrastructure, lack of mortgage agents and loan offices, underdevelopment of appraisers institution, etc.;

- high state dues, levies and taxes related to mortgage loans granting and issuing of mortgage bonds;
- passive state role in developing mortgage lending.

In Russia mortgage-baked securities (MBS) represent a type of asset-backed securities. Asset-backed securities allow their holders to collect revenue from a certain pool of assets. For the issue of MBSs the asset pool consists of mortgage loans receivables secured by immovables.

The Russian Federation's Federal Law "On Mortgage-Backed Securities" of October 14th 2003 lays down that there are two types of mortgage-backed securities: mortgage-backed bonds and mortgage participation certificates.

Mortgage-backed bonds are secured by pledge of the mortgage pool.

Bank refinancing through issue of mortgage-backed bonds on the security of mortgages takes place in the following way:
- A commercial bank concludes credit agreements with legal entities and (or) individuals secured on the immovables and executes respective mortgages;
- Mortgage agencies pay off mortgages at the bank through issue of bonds which are floated into the sock market;
- Cash assets gained from floating the securities return to the creditor;
- The state guarantees bonds interest payments.

Mortgage participation certificate is a registered security certifying the share of its holder in common ownership of the mortgage pool, the right to demand proper trust management of mortgage coverage from the certificate issuer, the right to collect money paid in fulfillment of the obligations that constitute the mortgage pool and other rights provided by the Federal Law. This type of security is absolutely new to the Russian market, and it implies a more complex financial mechanism. Each security purchaser has the share of the common ownership of the property constituting the certificates' "pool". A holder of such certificate is a kind of a mortgage loans owner and simultaneously acts as a settler and a beneficiary under the trust agreement. The holder gets almost all the money paid by the borrower minus compensation to the manager.

Mortgage is an indispensable element of a socially-oriented market economy. It attracts extra-budgetary investments into housing construction, provides effective redistribution of immovables ownership and expands the social base of civil society.

From the perspective of institutional theory the essence of mortgage lending may be explained in various interconnected aspects.

Mortgage may be regarded as an economic institution, a special tool for assigning functions to economic entities entering into credit relations.

Mortgage is an economic way of giving the status of a borrower on the basis of long-term contractual assignment of agreement fulfillment to subjects. This results in increasingly pronounced role of the institutional component in the rational economic behavior of the subjects.

Besides, mortgage may be considered an institution if the latter is interpreted as a system of formal and informal norms regulating the operation and the interaction of contracting parties concerning granting of mortgage loans and fulfillment of the obligations arising from them.

Housing construction is the foundation for mortgage institutions and mechanisms development. It is modernization, technical re-equipment of construction resource base and promoting of advanced technologies that should be the first step toward institutional transformation in the mortgage sector. Neglecting the leading and defining role of constructing branch institutions in developing mortgage may entail redistribution of budgetary resources to the disadvantage of the real economy. If the resources are committed solely to developing a network of specialized mortgage agencies, speculative trends in the market will grow, and banks and other investment institutions will derive revenues from exorbitant rates.

Economic models must develop in a balanced way. The economic modernization in Russia today predominantly pursues the improvement of its social parameters, shaping of civil society, consolidation of different population groups, overcoming of tensions and mutual estrangement between them.

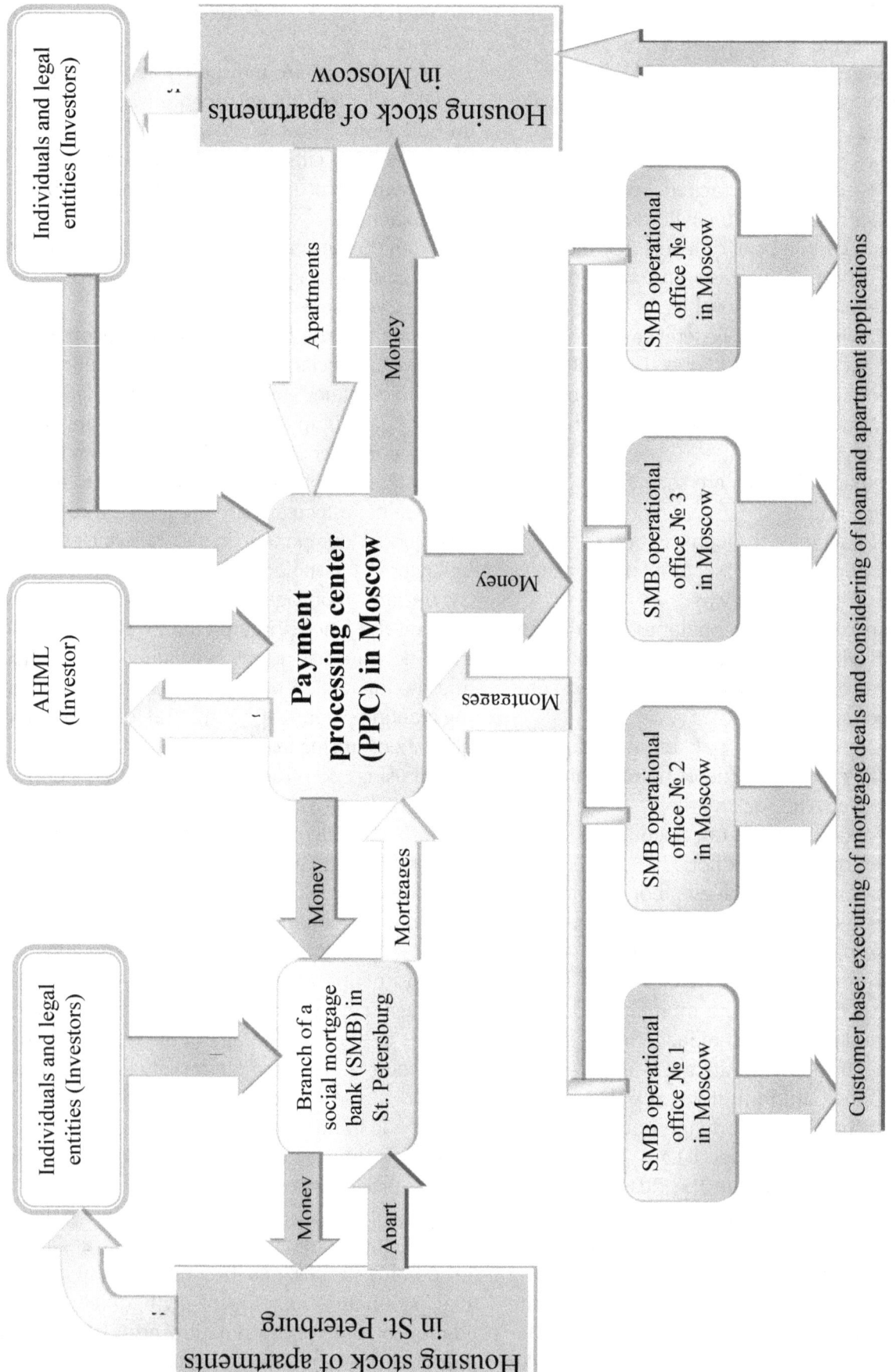

FIG. 1. Control structure diagram for the innovative initiative «Social mortgage bank»

In this context the system function of mortgage lies in social harmonization through mass satisfaction of a basic human need for housing via the market mechanism based on fair transactions controlled by the state.

The author is firmly convinced that when devising social mortgage innovations one should proceed from the assumption that social equity can be viable only if 100% of the population of a given country are provided with housing. This fact should not be underestimated. However, the present situation in Russia compares poorly with developed countries. Consequently, it is necessary to study and compare potential borrowers (for instance, the German with the Russian ones) before adopting the mortgage development model of a certain country. Evidently, their social standing, welfare, employment and earned income differ.

Therefore, it is necessary to distinguish between social natures of mortgage deals. Some borrowers have to be fully subsidized by the state for the whole crediting period if local authorities categorize them as socially deprived. Other socially important borrowers should be granted preferential interest rates for loans with due account for their formation and development.

The author maintains that mortgage in Russia is social in nature and in character of investments provided that the latter pay off by security sales rather than by overwhelming rates and commissions paid by borrowers.

In the medium term Russia needs strategic research support for the forming variety of mortgage institutions. Adjustment of American and German mortgage mechanisms in the Russian institutional environment is a nonlinear probabilistic process fraught with bifurcations and deferred negative effects.

There is a need for a sound comparative analysis of functions, structure, evolution, long-term positive and negative effects of various mortgage tools. Only on this basis can conclusions be drawn as to the prospects of their transplantation and co-functioning.

References

1. Zhukova, O. V. Innovative Systems of Management in Mortgage Based on European Construction Companies Practices. The international research-to-practice conference "Theoretical and Applied Aspects of Scientific Research" (January 30th 2015). Keynes Laboratory of Applied Economic Research. Collection of articles, pp. 20-26.
2. Zhukova, O. V. Non-bank Deposit and Credit Organizations as an Implementation Mechanism for Innovative Initiatives on Mortgage. The Journal of Regional Economic Issues №1(22) – 2015, pp.19-25.
3. Zhukova, O. V. Institutional Development of Residential Mortgage in View of International Integration Processes. The Journal of Regional Economic Issues №1(22) – 2015, pp.49-56.
4. Zhukova, O. V. The Lessons of the American Mortgage Crisis. 8th Annual research conference for students and post-graduates "Efficient Use of Regional Resources" Korolev Institute of Management, Economics and Statistics, March 20th 2008.

How to Be a Good Manager: Basic Behavioral Competencies

Yana Kochkova

Ulyanovsk State University, Ulyanovsk, Russia

Abstract: *In this article we describe the ideal model of behavioral competencies for managers. We also check presence of these competencies of managers (master's degree) at our university.*

Keywords: *management, behavioral competencies, expert methods.*

This topic is relevant because requirements for managers are growing in today's changeable market conditions. Besides the basic functions that must be implemented by managers, it is also necessary to create a list of the behavioral competencies needed. Competence can be defined as the personality characteristics that can help achieve the best management results.

In the current study, we defined a set of key behavioral competencies that modern managers should have. We constructed an ideal model and then conducted analysis aimed at identifying the current level of behavioral competencies of managers (master's degree) at Ulyanovsk State University. To determine which behavioral competencies to include in our model, we decided to interview experienced company executives from the Globe candy factory to get their thoughts on this subject.

During the interviews, we identified the following competencies:

1) Purposefulness
2) Responsibility
3) Communication skills
4) Stress resistance
5) Innovation and creativity
6) Common sense
7) Authority and leadership
8) The ability to see the future
9) Ethical behavior
10) The ability to make risky decisions
11) Initiative

It is important to note that these competencies do not contradict each other. Next we asked the managers to evaluate the degree of importance of these criteria to determine the ideal model using ranking. This data is presented in Table 1.

We used Kendall's coefficient of concordance to measure agreement between raters. It indicates the degree of association of ordinal assessments made by multiple appraisers when assessing the same samples [4]. Kendall's coefficient is commonly used in attribute agreement analysis. The formula is:

$$W = \frac{12 \cdot S}{m^2(k^3 - k)}$$

where k = number of factors and m = number of experts. Consequently, we get:

$$W = \frac{12 \times 9924}{13^2(11^3 - 11)} \frac{12 \times 9924}{13^2(11^3 - 11)} = 0.53.$$

The concordance coefficient can vary from 0 to 1. If it is significantly different from zero (W ≥0,5),

TABLE 1. Rankings of competencies

Competencies	Ranking by Experts													E	Deviation	Δ²
	1	2	3	4	5	6	7	8	9	10	11	12	13			
	Rank															
Purposefulness	1	3	2	2	2	1	1	4	4	1	1	5	1	28	-50	2500
Responsibility	4	1	3	1	4	3	6	1	5	3	2	6	4	43	-35	1225
Communication skills	11	10	4	9	8	9	8	5	6	9	7	7	9	102	24	576
Stress resistance	10	9	5	3	7	10	11	7	1	8	10	1	7	89	11	121
Innovation and creativity	2	5	1	4	1	2	3	6	2	7	8	2	11	54	-24	576
Common sense	3	4	6	5	6	8	2	2	3	2	3	4	2	50	-28	784
Authority and leadership	7	2	9	6	5	4	4	3	9	4	6	3	5	67	-11	121
The ability to see the future	6	7	8	10	3	11	5	10	10	6	5	8	3	92	14	196
Ethical behavior	5	6	10	7	11	7	9	9	7	5	4	10	6	96	18	324
The ability to take risks	9	11	11	8	10	6	10	11	11	10	11	11	10	129	51	2601
Initiative	8	8	7	11	9	5	7	8	8	11	9	9	8	108	30	900
Total														858		S=9924

it can be assumed that there is some agreement between the views of the experts.

So we obtained an ideal model Table 2) based on the ranking.

TABLE 2. Ideal model

№	Behavior Competencies
1	Purposefulness
2	Responsibility
3	Common sense
4	Innovation and creativity
5	Authority and leadership
6	Stress resistance
7	The ability to see the future
8	Ethical behavior
9	Communication skills
10	The ability to take risks
11	Initiative

With the help of expert methods, we evaluated the presence of these competencies at Ulyanovsk State University. We conducted a survey among master's degree students in management. A summary of the data from the survey is presented in the summary table below:

TABLE 3. Comparison table

№	Ideal model	Value	№	Received data	Value
1	Purposefulness	65	1	Initiative	62
2	Responsibility	65	2	Authority and leadership	61
3	Communication skills	65	3	Ethical behavior	60
4	Stress resistance	65	4	Communication skills	57
5	Innovation and creativity	65	5	Innovation and creativity	56
6	Common sense	65	6	Common sense	54
7	Authority and leadership	65	7	The ability to take risks	48
8	The ability to see the future	65	8	Responsibility	43
9	Ethical behavior	65	9	Stress resistance	42
10	The ability to take risks	65	10	Purposefulness	30
11	Initiative	65	11	The ability to see a future	29
E		715			542

As shown in this table, our model of competencies was ideal. The total value of this model was 715 points. The level of competencies at Ulyanovsk State University falls short of the ideal model (deviation - 24.2%). In principle, this deviation is small, but there are fundamental differences from the ideal model.

The first such difference is for purposefulness. In our opinion, this is the main competency for managers, as it is related to the ability to achieve goals. Without it, managers are not able to do their job. The work of managers consists of 5 processes, one of which is setting and achieving goals. We believe that the university must pay more attention to developing this competence in students. First and foremost, it is necessary to introduce discipline in the university, because it defines a lifestyle. Also, students need to know how to measure capabilities and time resources for businesses. And the most important thing is to teach them to set long-term and short-term goals and help them to achieve those goals.

According to the ideal model, responsibility should also be a key competency. Managers should be responsible for their actions, because if they take responsibility for themselves, they can be responsible and for others, they will trust the managers.

Thus, we can say that if these competencies are objectively necessary and in demand, the university can provide for their formation directly in the educational programs of higher and postgraduate education, or create an educational product for corporate universities and organize the promotion of this product in this segment of the market. In the latter case, education alliances must be formed with the corporate universities of different companies. It should also be noted that the subject of such interaction would be not only short-term programs, but also higher education programs, including the MBA.

References

1. Abaev, LC. 2012.Expert ranking of alternatives in the problems of big dimensions. In: (Name: Wilkins, al, editor). Managing Large Systems: the Collection of Works. 40: 19–29. (In Russian)
2. Korobov, VB. 2013. Some problems of practical application of expert methods. Scientific Dialogue. 3: 94–107.
3. Tambiev, SG. 2010. Competency approach to improving manager qualifications. World of Science, Culture anEeducation 6: 66–69.
4. Tolstova, MS, Hlopova, LA. 2012. Managers and their characteristics. Concep. 9: 54-58.

«Resource Curse»: the Subject of Discussion or Inevitability of Search for Future Energy Resources?

Kambar Musabekov, Shavkat Sunakbayev

International Kazakh-Turkish University named after K. A. Yassawi, Turkestan, Kazakhstan

Abstract. *This paper evaluates a situation concerning the most actual problem of the modern times - negative influence of natural resources plenty: as the Dutch disease, exhaustion of natural resources, degradation of power institutes and decrease in stimulus to accumulation of human capital.*

Keywords: *resource curse, Dutch disease, world economy, oil recovery, brain curse, clean energy.*

1. Introduction

Presence of mineral-raw material resources is the important factor of global competitiveness. The most intriguing empirically studied problems, actively discussed in the modern scientific literature on problems of economic development, include so-called «natural resources curse». Its essence ☒ the fact that in countries with plenty of natural resources rates of economic growth lag behind the same of countries with poor resources. Attempts to explain this phenomenon have led to revealing of some ways of negative influence of natural resources plenty on economic development. They include the Dutch disease, exhaustion of natural resources, possibility of degradation of power institutes and decrease in stimulus to accumulation of physical and human capital.

2. Method

The materials presented in this paper are based on the information from the books, scientific and statistical journals, laws and regulations of the regional and municipal authorities. The statistical data were collected from reports of the Department of Statistics of the Republic of Kazakhstan.

3. Problems

According to research of tendencies of development of many developed countries with plenty of natural resources, rates of their economic development are low enough and demonstrate inverse dependence on scales of resources at growing inequality of income of different social strata. Such negative relationship between natural resources plenty and economic growth has been named «resource curse». The term «resource curse» is a recent phenomenon, used for the first time by R. Auty in 1993 for description of the situation when countries with plenty of natural resources are unable to use their richness for development of own economy and demonstrate lower economic growth in comparison with countries with smaller scale of natural resources [1]. Such conclusions quite enough correlate with results of historical analysis of economic growth in the 17th century. Netherlands with their poor enough resources has overtaken rich Spain with precious metals, and at the beginning of the 20th century Japan has outstripped Russia.

However, the concept underlining that natural resources can be country curse, rather than advantage, has started to arise in 1980s yet. In different works, including well known analysis of J. Sachs and A. Warner, interrelation between plenty of natural resources and weak economic development of corresponding countries [2] is clearly traced. It seems that higher income on natural resources must give powerful impetus to economic growth. However, in real conditions such conclusion is wrong.

Research of models of development of countries with rich mineral-raw material resources shows that natural resources are not preconditions for successful development of national economy. One of the brightest examples of complex influence of resource abundance on economic growth - oil-producing countries. So, for example, in 1965-1998 GNP (Gross National Product) gain per capita in the OPEC (Organization of the Petroleum Exporting Countries) countries has decreased (on the average) to 1.3%, whereas in other developing states it was equal (on the average) to 2.2%.

The hypothesis stating that «resource curse» threatens, first of all, to countries with poor market development is confirmed by experience of development of countries with rich resources. 30 years ago two oil-producing countries - Indonesia and Nigeria had approximately equal GNP per capita, but by 2003 income per capita in Nigeria was at the level of 30% of the Indonesian. For the last 30 years the number of people living in absolute poverty in Nigeria has doubled and reached the mark 70%. Nigerians treat oil as a real curse, and this is taken into account the fact that Nigeria is one of the world's largest oil-producing countries - a member of the OPEC [3].

However, not all countries are subjected to «resource curse». Some of them, for example, Norway, Australia and Canada, demonstrate stable and high rates of economic growth and social development. Norway, rich with oil, is in the Top list of the index of human development of the United Nations. More over, the mentioned countries have shown efficient investment strategy, permanently investing significant capital in industrial infrastructure and development of human capital. Representation of various social strata in administration and state management provided economy transparency, created specific preconditions for implementation of efficient state economic policy.

The other possible variant of influence of natural resources on economic development - so-called «Dutch disease». The Dutch disease represents itself an economic phenomenon at which large income on export of natural resources leads to growth of budgetary income and induces increase of state expenditures (basically on social development), negatively influencing development of other sectors of economy.

Satisfaction of growing global demand in natural resources in this and the next centuries significantly depends on a concrete strategy, chosen by the world economic policy. Today the world community uses energy in enormous scales, and volume of power consumption increases catastrophically. This problem is aggravated by dynamics of growth of population of the Earth. In the 20th century at population growth in 3.8 times 15-fold increase in consumption of power resources was revealed. If such tendency remains, by 2030 world demand for energy resources will be 1.5 times higher the current value, and 45% of that demand is generated by China and India.

Nowadays threat of exhaustion of easily accessible hydrocarbon resources is obvious, significant growth of the world prices on energy resources (especially oil) is also evident. Oil still provides about one third of world power balance. In spite of large-scale development oil fields peter out, oil recovery decreases, environmental pollution in oil extracting areas promptly grows. According to some estimates in 20 years world oil consumption (in comparison with present volumes) will increase by 40%. Enormous growing volumes of extraction of organic energy carriers within the current century can lead to exhaustion of these resources. It is well known that fossil fuel resources, as a rule, are nonrenewable, and that creates a problem of long-term sustainability of economy of resources-rich countries. The world economy in whole spends for accumulation approximately 20-24% of its GDP (*Gross Domestic Product*), and for development of energy resources only 1-1.2% of GDP (or about 5% of volume of world real investments) [4].

The world energy crisis of 1970s, when humankind faced the fact of serious exhaustion of deposits of fossil fuel, stimulated search for new, alternative energy resources. According to an opinion of international experts, real energy revolution will start in the second quarter of our century, it will be characterized by radical change of structure and primary energy resources, advancement of strategy of renewable energy resources. Exactly due to wide application of renewable energy resources a share of consumed oil and other fossil fuel will be reduced.

4. Result

One of ways of negative influence of «resource curse» is connected with laws of development of political and economic institutes of different countries. It ruins, first of all, those states where such institutes are not developed. One of important conclusions underlines that «resource curse» is not imminent destiny of all countries rich with natural resources, but it threatens only to those from them where there is no real industrial strategy, where correct macroeconomic policy, allowing efficiently to manage income on resources extraction, is not applied. In countries rich with natural resources an important element of retention of power is redistribution of riches in favor of certain privileged social strata, rather than implementation of efficient economic policy. Huge income on natural resources feed up political corruption. Governments in such countries feel smaller necessity for formation of specific institutional structures, regulating national economy out of a fossil-fuel extraction sector, as a result, other sectors of economy significantly lag behind in development [5].

The other possible negative effect of «resource curse» is pushing out of human capital from such countries. Deficiency of human capital and low rates of its accumulation are considered by many researchers as major negative factors of «resource curse», influence of which is estimated at the level of 11-25%, depending on analyzed regions. Modern scientists consider human capital as the basic motive power of economic growth. Human capital provides permanent return on investments, while investments into physical capital are characterized by decreasing return. Higher educational levels allow to generate more knowledge and innovations, to facilitate adoption of technologies, to stimulate scientific and technical progress. A good indicator of dependence of economy from natural resources - a share of natural capital in national wealth, and volume of natural capital per capita can be treated as an index of resource richness. Analysis of data on 108 countries shows that resource dependence negatively influences accumulation of human capital [6].

Countries, relying upon export of natural resources, can neglect development of education, since they do not need it right now. On the contrary, countries with poor resources, for example, Taiwan and South Korea, put huge efforts for development of education, and that is one of main components of their economic success. The other key example - economy of countries of South East Asia. In 1960 GDP per capita in Singapore and Jamaica was approximately the same (2500 dollars). But Singapore started to invest actively in human capital. And Jamaica - in development of tourism. As a result, GDP per capita in Singapore in 2011 was equal to 50,700 dollars, and in Jamaica - in 10 times less. The most dynamically developing country in this region – Malaysia - has achieved tenfold increase in a level of living of citizens less than for 20 years. At low national education levels resource richness often leads to corruption growth, degradation of structure of economy and lower rates of economy growth, though first time it can quickly increase public expenses per capita. All these facts prove once again that a major factor of development of any country - not resources, but human potential, knowledge, physical energy. According to points of view of many researchers, if by the time of discovery of deposits in a particular country a high educational level of population is present, it is possible to say reliably enough that income, received on extraction of resources, will be used for the benefit of all people, accelerating rates of growth of economy and improving state prosperity.

In contrary to widely spread opinions, results of such analysis do not prove that countries rich with resources will live better if they get rid of such «preference». That is, «resource curse» is negative influence of structure of economy on rates of economic growth, not on a level of development. To develop faster, it is necessary not to destroy natural resources, but to replace them with other products.

«Resource curse» testifies to negative influence of not presence of natural resources, but their domination in national economy. Many countries with significant volumes of natural resources reach high human living standards and industrial progress. It is enough to mention the United States of America (USA), Canada, Norway with various kinds of natural resources. During more than two decades from the moment of establishment of «oil embargo» in 1974 GDP per capita (the best economic index of a level of living) in the OPEC countries decreased on the average by 1.3% per year, whereas in other developing countries it grew (on the average) with rate more than 2% per year [7].

5. Conclusion

Thus, «resource curse» is not just natural curse, induced by presence of plentiful natural resources, but (to some extent) «brain curse», thoughtless, infinite, uncontrolled consumption of natural resources. In spite of existing factors, contradictory to basic economy laws, general tendency of development of world economy for the last fifty years shows that the majority of countries rich with natural resources cannot reach higher rates of growth only by intensification of their use or by any other economic methods in comparison with countries not endowed with resources.

The future of any country is closely connected with alternative energy resources. The earlier humankind understands it, the brighter its horizons. In this connection it is reasonable to mention words of the USA president Barack Obama: «Country, which in the 21st century becomes a leader in clean energy production, undoubtedly will be a leader of global economy».

References

1. Richard, M., 1993. Auty Sustaining Development in Mineral Economies: The Resource Curse Thesis. — London: Routledge.
2. Sachs, J.D., 1995.Warner, A.M. Natural resource abundance and economic growth. // NBER Working Paper 5398.
3. Kondratyev of V. Mineralno raw material resources as factor of global competitiveness "A resource damnation" opportunity or inevitability?//World economy and international relations.-2010.-№6. - Page 20-30.
4. Grigoriev, L., Hooks of Century. World power at the intersection of roads: what way to choose Russia?// Economy questions. No. 12. - 2009. - P. 22-26.
5. Damania, R., Bulte, E., 2003. Resources for Sale: Corruption, Democracy and the Natural Resource Curse / Univ. of Adelaide.
6. Vasilyeva, O. Accumulation of the human capital and abundance natural ресурсов.//economy No. 12 Questions. 2011 of Page 66-76.
7. Gylfason, T., 2000. Natural resources, education and economic development// CEPR Discussion Paper 2594.

State Support for Small Businesses in the Chechen Republic

Muslim L. Tamaev

Chechen State University, Grozny, Russia

Abstract. *This article analyzes the current system of state support for small businesses in the Chechen Republic, with a focus on its effectiveness, problems of the system, and identification of measures necessary to improve its effectiveness.*

Keywords: *economy, state support, entrepreneurship, small businesses.*

Sustainable economic development, the development of social stability, and solutions for complex social and economic problems all require systemic and comprehensive support for small and medium enterprises (SMEs) at the federal, regional and municipal levels. In current economic conditions in Russia, and in particular in development in the Chechen Republic, SMEs are identified as one priority area. In this regard, further development and improvement of the whole system of state support of small businesses becomes necessary.

The government of the Chechen Republic has taken certain measures to support and develop small and medium businesses.

A state program currently being implemented is Support and Development of Small and Medium Enterprises in the Chechen Republic, 2014–2018. The plan is to allocate more than 1.2 billion rubles from the national budget. This program involves two subprograms: Support and Development of Small and Medium Businesses in the Chechen Republic, and Ensuring the Implementation of State Programs in the Sphere of Small Business and Entrepreneurship. Budget appropriations for the program from the state budget for 2014 amounted to 248.8 million rubles.

In addition, several nonprofit organizations have been created: Guarantee Fund of the Chechen Republic, Fund for the Support of Small and Medium Enterprises of the Chechen Republic, and Microfinance Fund of the Chechen Republic, which provide small businesses with support, such as the provision of guarantees, grants and subsidies.

The Guarantee Fund of the Chechen Republic granted 29 guarantees to small and medium enterprises in the amount of 115.4 million rubles. The Foundation for the Support of Small and Medium Enterprises of the Chechen Republic provided 149 loans in the amount of 84.8 million rubles. And the Microfinance Fund of the Chechen Republic issued 193 loans totaling 121.9 million rubles.

As of January 5, 2015, 22,700 individual entrepreneurs and 5,200 small enterprises were registered in the Chechen Republic.

Office centers (incubators) for small businesses have been built. These centers provide educational, consulting and legal services to entrepreneurs, provide office space on favorable terms, and offer a variety of financial services (micro-loans and grant support) for business development. The country currently has 10 such business centers: the Republican business center in Grozny; the production and business incubator in Grozny; business centers in Gudermes, Shali, Shatoi, Shelkovskaya, Urus-Martan, Argun, and Naurskaya; and a production rehabilitation center in the Znamenskoye village.

There have also been previous efforts to carry out regional programs to support and develop small and medium businesses:

- Support and Development of Small Business in the Chechen Republic, 2006–2010: "From Survival TO PROSPERITY!"
- Support and Development of Small and Medium-Sized Businesses in the Chechen Republic, 2011–2013
- Republican target program: Involvement of Youth in the Chechen Republic in Business, 2012–2014.

Key economic performance of small businesses in 2014 (end of year)

	Small businesses Total	Microenterprises (of total)
Number of active entities	5653	5551
The number of enterprises per 10,000 population		
Average number of employees (internal only)	20651	19293
The average number of foreign pluralist, man	163	163
Average number of contract employees, under civil law	243	123
The turnover of enterprises (rubles)	65.25 billion	61.12 billion
Investments in fixed capital (rubles)	10.23 billion	674.2 million

The turnover of small and medium enterprises by year amounted to: 2012, 30.4 billion rubles; 2013, 45.5 billion rubles; and 2014, 48.1 billion rubles. The share of turnover of small enterprises in the region's gross domestic product amounted to: 2012, 29.3%; 2013, 40.8%; and 2014, 40%. Thus, it can be concluded that compared with 2012, the share of small businesses in the GRP increased significantly in 2013 and 2014.

There are a number of key factors negatively affecting the development of small and medium businesses in the territory of the Chechen Republic:
- low social activism of entrepreneurs
- problems of staffing and training for small businesses
- inflation
- low development of mechanisms of state support of small business
- overgrowth of barriers to starting businesses (registration, licensing, certification, etc.)
- inaccessibility of credit resources for small businesses in commercial banks (Sberbank, one of the best options if based on its place in bank ratings, has lending rates up to 19.5 , depending on the form of credit).

The solution to these problems is the most important task of the state authorities at all levels, civil society organizations, business associations and entrepreneurs.

In order to solve the problem of credit resources, the Order No. 149-r of April 13, 2010 of the Government of the Chechen Republic created the Guarantee Fund of the Chechen Republic, which just in 2013 issued 35 guarantees for small and medium enterprises to attract credit funds.

The problem to be solved first is the tax burden. In accordance with the plan for priority measures for the sustainable development of economic and social stability at the federal level in 2015 (approved by Decree of the Government of the Russian Federation on January 2, 2015), adoption of legislation was suggested to grant subjects of the Russian Federation a reduction in the tax rate and introduce a "two-year tax holiday" for newly registered individual entrepreneurs in the field of industrial and household services.

Foreign and domestic experience shows that the creation of an infrastructure for support of small and medium businesses at the regional level contributes to the sustainable development of such businesses. A series of measures has been proposed with the aim of addressing the problems identified, and providing support for small and medium businesses at the regional level:

- optimization the legal framework for business, including tax legislation, to stimulate the development of the SMEs
- development of progressive financial technologies
- creation of a modern infrastructure of facilities to support small and medium enterprises;
- changes in existing programs for support and development of small businesses
- making the economic development of small businesses and measures to support them a priority, taking into account the level of socioeconomic development of regions and municipalities
- assistance in the creation and organization of public unions and associations of entrepreneurs

Implementation of these measures will provide substantial support and ensure the conditions for the development of small and medium businesses, as well as the development of municipal districts, urban districts, and sectors of the economy.

In short, it should be emphasized that the successful development of small and medium businesses is only possible through the political will of the state to create enabling social, economic, legal, and political conditions.

References

1. http://chechenstat.gks.ru/ [Accessed day/month/year].
2. http://chr.pmp.gkr.su/registry/program/ [Accessed day/month/year].
3. www.ChechnyaTODAY.com [Accessed day/month/year].
4. http://www.investchechnya.ru/maloe-predprinimatelstvo.html [Accessed day/month/year].
5. www.ChechnyaTODAY.com [Accessed day/month/year].

Education

The Role of National Upbringing in Formation of National Ideology of the Youth

Umida Boyzokova

Namangan State University, Namangan, Uzbekistan

A number of reforms are implemented in the Republic of Uzbekistan; they are intended to the modernization of the state and development of the democratic processes. Thereby the gradual increasing role of the local governmental power in the social, political and cultural life of the people is observed.

During the years of the Independence the state gained quite wide historical experience in particular regarding the formation of the state representative authorities through the formation of the national state basis, the implementation of a number of certain and gradual reforms under the democratization of the government. The gradual development under the democratization and liberalization of the society and the formation of new systems of the Uzbek statehood leads in its turn to the studying of the gained historical experiences and their scientific analysis.

Since the time the Republic of Uzbekistan had gained the Independence, it made a difficult and responsible way of the formation of a democratic state and civil society. As the President of the Republic of Uzbekistan I.A. Karimov notes in his book "Uzbekistan on the threshold of Independence": "our way to the independence and its development is not a road covered with flowers. The way we go is long and difficult; we try to liquidate the totalitarian hangover, to clean it off, to cure its disadvantages and damage done by it. Just therein lies the step towards the global civilization, the right way to the political and economic development" and just on this way "the in-depth searching and analysis of it doesn't lose its importance".

Nowadays under the intensive development of the globalization the protection of the consciousness, mind and world view of the youth against some negatives as well as the adoption of proper strategic direction of the upbringing and indoctrination of the national idea are of great importance.

The national upbringing is a direction of the practical activities, which is intended to the indoctrination of national traditions and legacy to the youth, to the formation of the folk mind in the world view, as well as to the recognition of own personality, strengthening of the national honor and imbuing the minds of youth with patriotism.

The national upbringing, as well as any other way of upbringing, has its certain goals. The national upbringing is begun at home by the parents and is continued at school by teachers, as the spirituality of every person, his/her world view, self-comprehension, as well as belief is formed by the family at home. In this respect the family is a true spiritual heart, an ideological factor and environment. In this basis the first concepts of the national ideology are indoctrinated by the family through the parental guidance and love.

The family as the environment, which keeps the sacred traditions, must inculcate the patriotism, faith, responsibility, motivation to train, as well as the cultural habits. No one could become an ideal human being without the consciousness of his/her own duties towards the family and motherland. The formation of the devotion to the motherland since childhood is of great importance. For this aim it is necessary among all the kinds of cognitive activities to choose those methods of upbringing, which have the folk mind. One shouldn't focus on foreign toys and movies; conversely we need toys, games, tales, and books, which have national oriental flavor. Thereby the mind of a child is filled with the respect to the national values and patriotism.

A representative of a nation, who has a developed national consciousness and well-formed world view, would exalt his/her nation, defend its interests, understand his responsibility towards his/her nation, would not exchange his/her interests to the good of the other ones; his/her devotion to the motherland would be constant. He/she would be proud of being the representative of his/her nation; his/her life would be filled with the consciousness that it is the supreme good and spiritual wealth for a human being. But the current process of the globalization has a negative impact on the national upbringing.

Today our way of life and all the spheres of our life are affected by traditions, habits and values of the other countries. Of course we can adopt those of them which are suitable for us, but the blind adoption of these foreign habits and values instead of our own ones is beyond the purposes.

For example we have a number of time-honoured values which are of the most importance for us: respect for the parents, filial piety, protection of the holiness of the family values, obligations towards the parents, obedience, collaborative planning of some important things, good fellowship between brothers and sisters, parents and children, children-in-law and parents-in-law, respect of a pupil to a teacher etc. But nowadays the impact of European individualism between the kindred increases, and sometimes it leads to that there is no one to follow a close one to the grave. This phenomenon becomes commonplace gradually and thus as a preventive measure it is of great necessity to pay a special attention to the implementation of the national upbringing.

The experience of Japan, which is one of the most developed countries of the world, could be in this respect quite demonstrative, because the national values descend from parents to children in this country over the centuries. They could not only achieve a great success, but also could keep their traditions thanks to the cohesion of the society. It is not a secret that along with the development of industrial management, the creation of new technical equipment and technologies and consolidation of their position on the world market they dodged their national upbringing.

Conclusions and motions:

— Awakening of will to work in the youth, upbringing of an active and eager person, who is able to self-teaching;

— Upbringing of a person, who has a national pride, responsibility, patriotism and confidence in his/her future;

— Formation of feelings of humanism, truth and patriotism in the youth;

— Upbringing of the developed youth, who show regard for the legacy of their predecessors.

References

1. Karimov I.A. Uzbekistan on the threshold of Independence. — Tashkent: Uzbekistan, 2011. — p.28
2. Karimov I.A. 16 years of independent development of Uzbekistan // The way of modernization and sustained development of the economics of the state. — Vol. 16. — Tashkent: Uzbekistan, 2008. — p.4.
3. Rasulova N. Youth wings of the political parties // Public opinion. Human rights. — 2008. — vol. № 4. — p.135–138.
4. "Enlargement of the youth participation in work of the institutions of the civil society relating to the effective forms and methods" Educational materials prepared for a number of regional educational seminars. — Tashkent, 2009.
5. http:/www.adolat.uz

Methods of Ecological Knowledge Training of Pupils and Celebration of the Festival "Day Of Birds" within the Framework of the Out-of-Class Activity

Sulgun Durdyeva

School for Gifted Children named after the hero of Turkmenistan A.Niyazov, Ashgabat, Turkmenistan

Abstract. *The article is directed to the formation of the ecological knowledge of pupils, training of the responsible and careful attitude towards the environment and all the living beings, as well as to the studying of nature of the native country. The questions of development of the conscious, responsible personal behavior of pupils, which is relevant towards the situations of reality, as well as the decision of problems in the course of practical activity on the nature protection, should become the major aspects of the school activity.*

Keywords: *nature, training, excursion, plants, reserve, desert, environment.*

The ecologization of training process and upbringing of the rising generation became nowadays one of the basic directions of the state policy of Turkmenistan in the sphere of education.

In the system of school education the biological disciplines take a special place, bringing an essential contribution to the all-around personal development and forming a modern natural-science worldview among the people of the rising generation. The organization of any out-of-class work provides for that within the limits of obligatory programs. Its content and forms depend especially on the interests and needs of pupils, as well as on the local conditions.

The biology teacher should pay a great attention to the organization and holding of the out-of-class biology activities. One should organize the round-tables for various biological problems, traditional festivals of harvest, birds, flowers, as well as one should gather with pupils the drug plants, seeds of tree and bushes, in the context of the organization of school and interschool competitions.

All of these measures can be taken thanks to carrying out of special work with the rising generation. The little citizens should realize since childhood that the problem of the environmental control and all the living beings, perhaps, is one of the most topical, and the task of the elder generation, leaders, and teachers consists in the teaching of thrifty usage of the nature.

However, how to make the out-of-class lessons interesting and meaty, how to involve the pupils in preparation and carrying out of biological measures? The teacher should take up here the running.

During the carrying out of any measure the teacher should define the nature of work clearly, make a plan of its carrying out and explain the task to the pupils who will perform it. There are such measures which preparation takes 5 or 6 months of the academic year, for example the festival of birds.

The out-of-class work has a great value during the studying of birds at biology lessons, because it develops the ecological culture of pupils, foster their care to birds and the nature in whole.

The populations of birds are being reduced due to the influence of the economic activities of human, and it is an integral part of the nature. What can be the best beauty of parks and gardens? Is it really possible to imagine the spring without sonorous bird's din and songs? Birds are our helpers.

In the initial classes the biology teacher can organize the observation over the nature, drawing up the calendars of weather and the nature. As a part of the observation it is possible to recommend to the pupil to take note of the arrival or departure of birds etc. Thus, the pupils of the initial classes will have a defined idea of the conducting observation over the nature by the transition to the upper school.

Nowadays we are faced with the problem of not only the preservation of nature and its reasonable use, but also with the problem of its further enrichment. All of these problems concern the birds' world, and they are much more difficult than it seems at first sight as compared to the problems of the national-economic value.

The celebrating of the festival "Day of birds" could bring up the love to the feathered tribe at children. During a huge work directed to the preparation of celebrating of the festival "Day of birds" the pupils make some «discoveries» for themselves, and it prepares them for the independent actions concerning the wildlife management. We believe that such holiday will necessarily help to bring up such humanities as kindness, sympathy, fidelity, responsibility which lack is observed among our children nowadays.

The organization and celebrating of the festival "Day of birds" at schools concern to the out-of-class measures. The large-scale participation is one of the main features of the festival. These are the first steps which could make children "green" and inculcate them the understanding of the importance of being in harmony with the nature and with all the living beings.

Tasks of the celebrating of the festival "Day of birds" and preparation to it

The celebrating of the festival "Day of birds" is based on the following problems:

1) training of love to the nature among the pupils;
2) attraction of useful birds of the city;
3) involving of pupils to the nature conservation activity concerning the bird protection;
4) propagation of knowledge on the nature conservation among the pupils and increase of their level of the ecological culture;
5) prevention of the actions which threaten the existence and development of the nature reserves and wildlife preserves in the matter of the protection of rare bird vanishing species in Turkmenistan.

All the pupils as well as the pedagogical staff should take part in carrying out of the festival "Day of birds", but the leading part in this measure play the pupils of the 5th, 6th and 7th grades. It is necessarily also to involve in the preparation of the festival the experts in the ornithology, scientists and non-governmental organizations for the nature protection.

The main attention during the festival and before it should be paid to the business part, that is the handcrafting and hanging of birdhouses, as well as the celebratory preparation (music, fancy-dresses, masks, quizzes, wits and humor competitions) which accompanies the abovementioned measures and adds the solemnity and beauty to the festival.

The following measures should be included in the preparatory work for the celebrating of the festival "Day of birds":

1. Handcrafting of birdhouses (nesting boxes, birdfeeders etc.).

2. Carrying out of roundtables, writing of summaries and reports about the value of birds, birds of the city or settlement, rare and vanishing species of the feathered tribe as well as about the ways of help them to survive.
3. Issue of special wall newspapers, posters and photomontages devoted to the feathered tribe.
4. Making of some visual aids and organization of exhibitions of hand-made crafts or drawings.
5. Working out of the musical program.

All these measures should be taken on the initiative of children and with a help of senior pupils. Biology teachers and instructors should educate the pupils before celebrating, and thereby should interest them in carrying out of this festival. Certainly they are to give line to children to improvise; especially it concerns the art part (for example, the verses, small shows, demonstration of films).

During the first years of celebrating of the festival the carnival processions were very interesting. Unfortunately it is a thing of the past. But this tradition can be revived in form of a competition for the best mask or fancy-dress representing any bird. It will also give a chance to include the girls in participation more actively.

Handcrafting of the birdhouses

It is assumed that handcrafting of the birdhouses is a waste action and a thing of the past. But the classrooms for the handicraft lessons are in each school and formerly the handcrafting of the starling nest boxes was a part of the education program for boys. We believe that a proposal to handcraft some starling nest boxes will be well received by the boys.

Handcrafting a birdhouse it is necessary to take into account the following:
a) Birdhouses are made of dry timber;
b) birdhouses are made only of boards, logs, stick, but not from veneer:
c) birdhouses should not have any ornaments and cracks;
d) the bottom of the nest should be inserted on the inside;
e) boards on the inside should not be planed;
f) the starling nest boxes should not be painted in bright colors:
g) the size of the entrance of the birdhouse should correspond to the size of a bird, for which this birdhouse is intended, the entrance should be placed in the upper part;
h) it is necessary to put a serial number on each birdhouse (this number will help in observation, as well as in the record at feed studying, reproduction of birds which colonize the birdhouses etc.).

Figure 1. A birdhouse

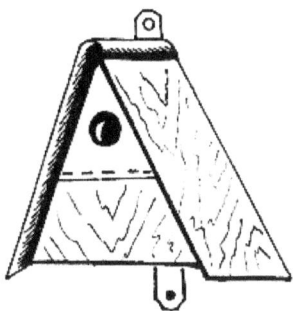

Attention

It is forbidden to nail up the bird houses to trunks and branches of trees. The nails are the reason of disease of trees because they damage wood of a trunk and lead to drying up and destruction of trees. The quantity of the bird houses necessary for hanging depends on the locality, delivery of feedstuff and other conditions concerning the biology of birds. It is necessary to remember that attraction of birds is connected not only with the presence of bird houses and feedstuff, but it depends also on quantity of bird houses on one tree. It is connected with habits of birds.

Within the framework of the ecological education those observations are considered to be the most valuable which make the pupil to draw conclusions about the value of living beings in the human life, to estimate their condition on the surveyed territory and generate a desire for independent labor directed to the improve of the environment.

Wide use of various tasks within the out-of-class work, as well as the observation and experiments develop the research abilities of the pupils. Besides that, the concreteness of the observable phenomena, the necessity to record the observed briefly or to draw corresponding conclusions, and then to tell about it during the lesson develop among the pupils such internalities as the thinking, observation, forces to think of the fact which called their attention earlier. The individualization of training is easily carried out and the differentiated approach is realized within the out-of-class work.

The purpose of the ecological education consists in the formation of the responsible behavior towards the environment which in its turn creates the basis of the new thinking and assumes the observance of the moral and legal principles of the natural management, the propagation of ideas of the optimization, the vigorous activity for studying and protection of the country, as well as the protection and renewal of natural wealth.

Role of Labour Education in the Ethnopedagogics in the modern world

Nailya Zaydullina

Naberezhnye Chelny Institute, KFU branch, Naberezhnye Chelny, Russia

The devolving of knowledge to the younger generation isn't any more a pledge of its successful integration into society in the modern promptly changing world. The today society is in need of well-educated, moral, creative people who can independently make critical decisions in a choice situation. Therefore the ability to orientate in a sea of information and ability to make the correct decisions on the base of data gained from the various sources turn into the key factors for creation of a personal vector of development.

The priorities of the modern education are defined in the final report of UNESCO, and it has to be said that they don't contradict the traditions of the Russian school at all. These priorities are the following:
- to teach to gain knowledge, that is to learn to study;
- to teach to work as to work to earn, that is the doctrine for work;
- to teach to live, that is the doctrine for life;
- and to teach to live together with the other people, which is often not the same thing as the previous one, that is the doctrine for the joint life [5].

The 18th article of the Federal Law "On Education" of the Russian Federation states: "The parents are the first teachers. They are obliged to lay the basis for the physical, aesthetical, moral, legal and intellectual development of the identity of a child at a preschool age". The family is the environment, where the child develops and which lays the basis of his/her personality. The family gives to the child the main things, which can't be given by the other social institutes; these things are the intimate and personal communication as well as the initial unity with the natives.

Within the family the child is being constantly educated by means of the authority of the adults, by their personal example, as well as by means of the family traditions. But there are no any accurate organizational forms within it. The child is educated through the activity of his/her family and by means of somebody's individual impact on him/her. Within the family the process of socialization of the child occurs, and it is incomparable in order of the educational importance, because it provides the children with versatile knowledge of the surrounding social reality, it introduces them the human culture.

However, according to the researches, the current main shortcomings of the family education consist in the ignoring of the maternal language, traditions, spirituality, history, native culture, as well as in the indifference of the family to the problem of the strengthening of the continuity of generations in the context of the national development [2].

In the late eighties of the 20th century the questions of the labor education were removed from the foreground. The concept of the preschool education mainly began to be related to the upbringing of a healthy, intellectually developed personality; according to the Soviet pedagogics the main idea of the education consisted in the fact that the education depended on the work of the person, i.e. every pupil was responsible for his/her knowledge. The new generation lacks obviously this sense of responsibility. The young people want to live the good life; i.e. their requirements are out of phase with their opportunities (that means in this case the lack of workplaces, the level of professional education, the ceasing of care for the professional education on the part of the state, absence of mentors). This problem is observed also in the sphere of the preschool education.

Researches of the last five years show that modern preschool children want to be bankers, models, to trade in the market in the future; the hairdressers are rarer. Many of them know modern car models, want to have them, but none of them want to be a driver. Polls of children in the recent years show that nobody wants to be an astronaut, a captain, or a teacher.

The problem of the labor education at the level of the preschool education consists in the obsolescence of the methodological basis of the theory of the labor education. And the reason of it is its eradication during the shift of the economic development of the country.

The cultural development of the regions and the country is in general ought to develop the complete viewpoint, attitude and outlook which are adequate to the reality and problems of entry into the Bologna agreements, having kept the basic native moral values and priorities at that.

Having entered the Bologna agreements it is essentially important for Russia to use own cultural capacity for the formation of a positive image of the country abroad.

Therefore it is necessary to pay a special attention to disclosure of the contents and features of the labor education in the national Tatarian pedagogics, of the role which a traditional Tatarian family plays in this process [4]. For many hundreds of years the family education at Tatars acted as if not the only, then as the main form of child education in the society. The special significance in the Tatarian family was attached to the labor education. From the earliest age the children carried out a number of labor instructions which became more and more complicated over the years. A growing-up child in his/her family was constantly focused on the work. The concentration of the children on the work wasn't only the realization of the consequence of any «pedagogical ideas» of their parents, though it had a deep pedagogical significance. It was severe need as the labor life of the adults was too heavy and it wasn't possible to manage it without the child labor. One can say that the life itself gave to the younger generation the labor education, not just the parents.

The Tatarian people always connected the labor education with the hope for the future and fairly considered the labor education to be a basis of moral education. People always dream and hope that the adult children will serve the parents just as they looked after them during their childhood. According to an ancient treatise, "A person must never remain without care and work, i.e. if you want to feel pleasure in your life, you ought to live by own labor, because there are no pleasures without an effort, and every work is followed by the pleasure" [4, page 12].

One of the thinkers of the period of the Kazan khanate was the poet Mukhammedyar (the end of the 15th — the early 16th century). His ideas of the labor education remained in his works «Tukhfai Mardan» («A gift of Dzhigits»), «Nasikhat» («Manual»).

Mukhammedyar sought a way to make the people happy in the real, terrestrial world. According to his statement, a person can become happy only thanks to his/her mind, creative force and energy, as well as free labor. The earth for Mukhammedyar was a place of the fulfillment of noble and good deeds after which the person had to strain.

According to him, people are destined to work wonders in this world for the sake of the happy life. As the poet considered, it was just the highest ideal of the mankind, which deserved a special eminence, praise and respect [4, page 14].

Although the ideas of Mukhammedyar are social and utopian, they have a historical and pedagogical value. Opposing the existing orders and believing in the justice victory he looked for ways of improvement of life of the people. "I believed in the human, I glorified the mind, I learned to appreciate the best human qualities". It is also remarkable that the word «human» in the works of Mukhammedyar is associated with the concept of a simple toiler. He eulogized such human qualities as bravery, courage, diligence, sang of honesty and humanity.

Mukhammedyar managed to show the moral superiority of the people of labor who worked honestly and therefore knew the price of the daily bread [4, page 19].

In the poem «Kutadgu Bilig» the poet Yu. Balasaguni discloses the wisdom of the ancient east pedagogical thought. The labor education of the younger generation can serve as a reliable reference point in the subsequent course of life of the child because this thought of the poet is time proved [4, page 12].

Qol Ghali (the end of the 12th — the early 13th century) is a bright representative of the poetry of the Bulgar period. The glory of labor was also sung in his well-known work «Kyssai Yusuf» («The legend on Yusuf»).

The period of work of Qol Ghali was marked by big labor feats. This time was characterized as the golden age of the town planning, because by the beginning of the 11th century the Volga Bulgaria had been covered by a dense network of cities and villages. There were over 200 cities on its territory.

The main character of the poem by Qol Ghali is called Yusuf and he is an ideal. He is always ready to work on himself, to work on changing of his character to the best, is ready to self-improve. Qol Ghali also does a view of him as an example for imitation for the younger generation, as a perfect and ideal hero. It seems that Qol Ghali emphasizes that there aren't people without any shortcomings, but the one who is capable to correct the shortcomings, strives for perfection, purity of the thoughts and actions is considered to be the best. The idea of the labor education takes one of the main places among the art esthetic views of the poet Qol Ghali [4, page 6].

The educators of the Tatarian people took up widely the questions of the labor education in their poetic works, disclosed them in an integrated manner, in combine with the other directions of the education.

The ethnopedagogics can be defined as an organic component of the general pedagogics and at the same time as an independent branch of the scientific and pedagogical knowledge in the system of the humanities which studies the regularities and features of the national education (in broad social sense); it is intended for purposeful reproduction of knowledge, abilities, skills, features of a personality which are valuable to the ethnos.

The scientific staff of the ethnopedagogics laboratory of the Research institute of the family and education of Russian joint stock company headed by the academician Volkov G. N. define the ethnopedagogics as the science which studies the national culture and national pedagogics for the purpose of detection of the general regularities of their formation and development as well as the opportunities of use in modern educational systems. The basic scientific concepts are the following:

— basic pedagogical concepts of the people (living, education, self-education, re-education, manual, training, schooling);

— child as object and subject of education (natural child, orphan, adopted child, age-mates, friends, other peoples' children, children's environment);

— functions of the education (preparation for work, formation of the morally and strong-willed features of character, mind development, care for health, instillation of love to the beautiful);

— factors of the education (the nature, a game, the word, communication, tradition, business, life, art, religion, an example — an ideal (persons — symbols, events — symbols, ideas — symbols);

— methods of the education (belief, an example, the order, an explanation, schooling and exercise, a wish and blessing, a spell, an oath, a request, council, a hint, approval, a reproach, an arrangement, a precept, a precept, pledge, a repentance, the sermon, the will, a ban, threat, a damnation, abuse, punishment, a beating);

— educational tools (counting rhymes, proverbs, sayings, riddles, epos, fairy tales, legends, myths, etc.);

— organization of education (labor associations of children and youth, youth holidays, public holidays) [4].

Thus, the labor education of the younger generation has to become one of the tasks of high priority; differently it can turn into one of the reasons of the extinction of the nation.

References

1. Zaydullina, N. N. Folklore traditions in labor education [Text] / N. N. Zaydullina//the Bulletin of the University of the Russian Academy of Education. — Moscow, 2007. — No. 3 (37). — Page 66–67.
2. Kozhanova M. B. Traditional values of a family within the context of education. [Text] / Kozhanova M. B. Ethnodidactics of the Russian people: from the national educational systems to the global space. Proceedings of the VIIth International academic and research conference. — Nizhnekamsk, 2009. — 45 pages.
3. Malikov R. Sh. Turkic-tataric humanistical pedagogical idea of the Middle Ages. — Kazan, 1999
4. Mukhametshin A. G. Development of the pedagogical idea and education of the Tatar people in 18th - early 20th century [Text] / A. G. Mukhametshin. — Naberezhnye Chelny: NGPI, 2008. — 138 pages.
5. Nezdemkovskaya G. V. Problems of the formation of the categorical definitions of the ethnopedagogics. [Text] / Nezdemkovskaya G.V. Ethnodidactics of the Russian people: from the national educational systems to the global space. Proceedings of the VIIth International academic and research conference. — Nizhnekamsk, 2009. — 17 pages.
6. Pichugin S. S. Revisiting the realization of the state standards of the second generation at the elementary school. [Text] / Pichugin S.S. Ethnodidactics of the Russian people: from the national educational systems to the global space. Proceedings of the VIIth International academic and research conference. — Nizhnekamsk, 2009. — 310 pages.

Mathematical modeling of production processes using MS EXCEL

Elena Mirzoyeva
Kuban State University of Physical Education, Sport and Tourism, Krasnodar, Russia

Abstract. *The article is concerned with construction and use of production functions for the purposes of modeling of production processes of an economic unit. The article presents the algorithm and mathematical methods used to construct production functions.*

Keywords: *construction of a production function, mathematical modeling of production processes, Excel Wizard.*

Production functions are used to model production processes of an economic unit: a separate company, an industry or the national economy as a whole.

Production functions allow to meet the following challenges:

- evaluation of return on the resources in a production process;
- economic growth forecasting;
- development of business plan options;
- optimization of operation of the economic unit under conditions of specified criteria and resource constraints.

Here is a general production function:

$$Y = Y(R_1, R_2, ..., R_i, ..., R_n) \quad (1)$$

where: Y — indicator used to characterize production results;

R — factor indicator of i production resource;

n — number of factor indicators.

Production functions are defined in the form of one-factor and multifactorial statistical relationships - regression equations.

Construction of a production function is based on a preliminary statistical data analysis. The initial data for the construction of the production function are given in Table. 1.

First stage – *data analysis by samples*.

Data analysis by samples consists of the following procedures (using Excel Wizard):

- determination of standard quadratic deviations (δ) of each of Y and R_i samples (Excel: statistical function STDEV);
- determination of sample variances (σ) (Excel: statistical function VAR);
- determination of medians, min and max of each sample (Excel: statistical functions MEDIAN, MIN, MAX);
- Z-testing of samples (Excel: statistical function ZTEST).

Second stage – *correlation data mining*.

Pair correlation coefficients (R) are calculated for this purpose (Excel: statistical function CORREL). They vary between -1 and 1. The closer the

TABLE 1. Initial data for the construction of the production function

Resulting indicator Y	Factor indicators $R_1\ R_2\ R_3\ ...\ R_i\ ...\ R_n$					
Y_1	R_{11}	R_{12}	R_{13}	... R_{1i}	...	R_{1n}
Y_2	R_{21}	R_{22}	R_{23}	... R_{2i}	...	R_{2n}
...	...					
Y_j	R_{j1}	R_{j2}	R_{j3}	... R_{ji}	...	R_{jn}
...	...					
Y_m	R_{m1}	R_{m2}	R_{m3}	... R_{mi}	...	R_{mn}

correlation coefficient value to 1 or -1 is, the higher the degree of correlation of respective random values is.

The same is done with factor indicators which correlation coefficients are very close to zero.

Third stage - *regression analysis*.

It is advisable to make sure in the right choice of factor indicators for the purposes of production function modeling before the construction of a regressional relationship. The F-test is carried out for this purpose (Excel: statistical function FTEST).

In addition, Excel standard functions LINEST and LOGEST provide an opportunity to obtain additional statistical characteristics of a regressional relationship.

The order of evaluation using LINEST function is as follows:

1) enter initial data or open an existing file that contains the analyzed data;

2) select an area of blank cells 5x3 (5 rows and 3 columns) to display the results of the regression statistics;

3) open the Function Wizard dialog box by either method:

a) select the **Insert/Function** in the main menu;

b) click the **Function Insert** on the **Standard** toolbar;

4) in the **Function category** box (Fig. 1) select the **Statistical**, in the **Function name** box - **LINEST**. Click on the **OK** button;

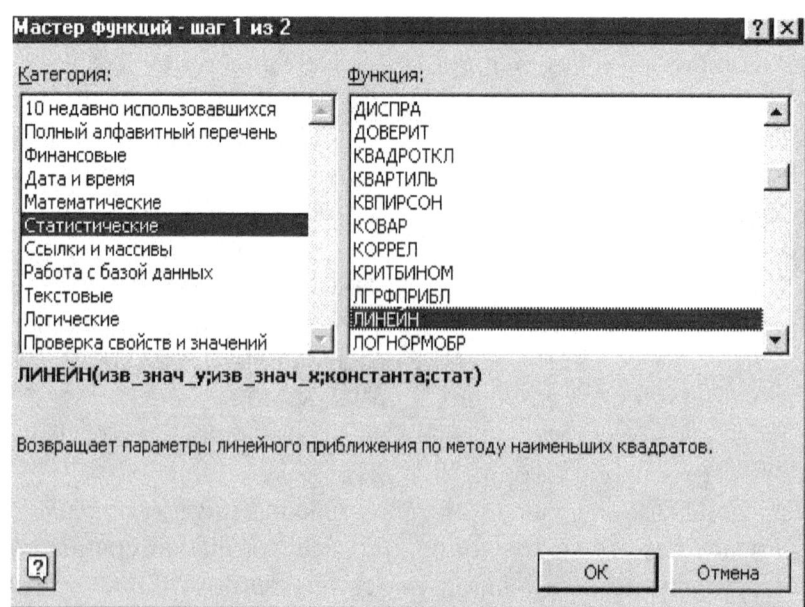

FIG.1 Function Wizard dialog box

5) type in function arguments (Fig. 2):

known_y's – the range containing resultant characteristic data;

known_x's – the range containing the data of the independent characteristic factors;

const – the logical value that points to the presence or absence of the absolute term of an equation; if *const* = 1, the absolute term shall be calculated in a conventional manner, if *const* = 0, the absolute term is equal to 0;

stats – the logical value that shows whether to type in additional information on the regression analysis or not. If *stats* = 1, it is required to type in additional information, if *stats* = 0, just estimates of equation parameters are displayed.

Click the **OK** button;

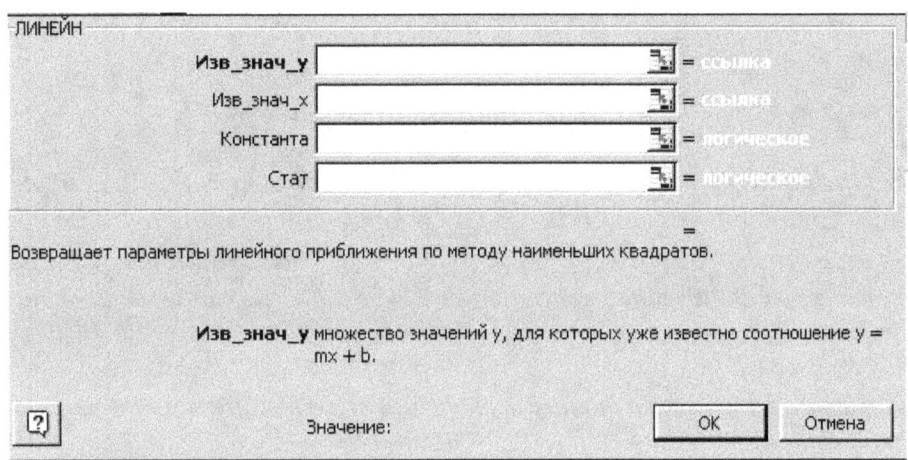

FIG. 2 Dialog box where LINEST function arguments are typed in

6) the first element of the final table will be displayed in the top left cell of the selection area. To expand the entire table press <F2> key and then the <CTRL> + <SHIFT> + <ENTER> key combination.

Additional statistical characteristics of the regressional relationship will be displayed in the order shown in Table 2.

TABLE 2. Additional statistical characteristics of the regressional relationship

A_2 coefficient value	A_1 coefficient value	A_0 coefficient value
Standard deviation A_2	Standard deviation A_1	Standard deviation A_0
Determination coefficient R^2	Standard deviation Y	
F-statistics	Degrees of freedom	
Regression sum of squares	Residual sum of squares	

It is necessary to remember that it is proposed to construct production functions in the following assignments for submission in the linear and power forms using Excel standard function LINEST from the statistical functions section:

$$Y = A_0 + A_1 R_1 + A_2 R_2 + ... + A_i R_i + ... + A_n R_n \quad (2)$$

$$Y = A_0 R_1^{A1} R_2^{A2} ... R_i^{Ai} ... R_n^{An} \quad (3)$$

where A_i — initially unknown regression coefficients.

References

1. Mirzoyeva E.V. Mathematical methods and models in economics: Textbook / E.V. Mirzoyeva, S.N. Gritsyuk, V.V. Lysenko – Rostov-on-Don: Phoenix, 2007- 348 p.

Quality Management of Education in Child Welfare Centers

Anatoly Red'ko

Musical College, Perm, Russia

Abstract. *The article deals with some current issues of child welfare centers (by the example of the Palace of Children (Youth) Creativity), and ensures the quality of supplementary education for children in the modern world. Since it is proved that the core of the quality management of art education in a municipal center of supplementary education of the younger generation should be the value-motivational approach, the article provides the basic principles of this approach and clarifies its nature.*

Keywords: *supplementary education, quality management of art education, value-motivational approach.*

After the enactment of the Federal Law "On Education" in the Russian Federation the special schools for gifted and talented children were simply forgotten. Nowadays in Russia there are no more than a dozen of such schools; although during the years of the Soviet Union the artistic elite was brought up there, and this elite still determines the leading role of our country in the world culture. The system of continuous art education is based on the fact that some creative predispositions of a child are detected at an early age and at the same time the gifted child gets the development of his/her creative abilities in an institution of such kind. Prior to the enactment the Federal Law "On Education" the students could receive the innovative integration training, but from the 1st of September, 2015 they won't have such opportunity. According to this law, up to the fifth grade "the vocational education is equated to the hobby groups. Schools should perform the functions of conventional educational centers themselves". [2] One must prepare them for it from the first grade.

In estimating the prospects for supplementary education of the younger generation one should emphasize that the familiar situation of satisfaction the curiosity for the state money becomes a thing of the past. All the municipal institutions are faced today with a necessity of implementation of market-oriented projects which could bring profits. Their task consists nowadays not only in the gathering and devolving of some new knowledge, but also in the formation of the entrepreneurial culture, technology commercialization, capitalization of results of intellectual and creative activity. We note the appearance of a new model of the supplementary education of the younger generation; this is the institution of business type and we call it the project-oriented one. Nowadays there are many approaches to the problem of formation of a new model of a supplementary child care institution. As a researcher I am anxious about the driving of these institutions into the Procrustean bed of vocational education, i.e. about the literalization of this principle as a single function, while ignoring the other aspects of education, which could result in its dehumanizing. Being a young scientist I told a lot in the last decade about the fact that the institution (i.e. the Palace of Children (Youth) Creativity) fulfills no longer its purpose, which consists in being a social center, whose aims are to accumulate and devolve the spiritual values. We incline to the opinion that the supplementary education of the younger generation is in the system target crisis. The previous values meet no longer the challenges of our time to the full extent.

The problematic issue consists in the simultaneous keeping the tradition with the responding the challenges of our time. Every modernization supposes a clear understanding of the ultimate goal. Making no questions of the importance of the invariant values of the supplementary education, we can't deny the presence of some specific problems at the same time. The world changes and the training can't be unchanged.

Centers of the supplementary education become more and more popular, and still remain one of the most popular centers of socialization.

Which current external conditions determine the framework for the development of the supplementary education? Do they really open up the new opportunities? Let's try to evaluate them objectively.

It is necessary to build a scientific and methodological basis for the development and implementation of new artistic profile programs.

A competence approach has been chosen as the conceptual basis for the supplementary education. It has been determined by the educational standard. In its new version the profile structure of programs is framable to the limit. Their composition is formed by the municipal educational institution itself, with a glance of the opportunity of diversified disciplines, modules, courses and practices. It is obvious that in our time another integrative, holistic approach to the development of these kinds of programs is required. The supplementary education of the younger generation is in need of interdisciplinary projects and researches. The educational standard ensures the right of choice as well as the structure and the content of the art programs. The result is fixed, but the path to it is determined by the teaching child care institution, using increasing opportunities of social partnership with science.

The combination of the disciplinal and the modular principles ensures the preservation of the tradition and provides at the same time the sensitivity and responsiveness to the necessities of the modern world. The module as a special element of the educational programs corresponds to the nature of competence, allows to harmonize theory with practice, to build interdisciplinary links, ensures the integrity of the professional world outlook of students, creates conditions for the application of knowledge and skills in using tools of one genre, to solve problems in other related genre, provides insight into the value and applications of skills in professional activities.

The higher is the degree of freedom, the higher is the requirements for the quality of the supplementary education of the younger generation as well as the need for its objective, reliable confirmation. As a result there are new management tools such as monitoring of the effectiveness of training activities, professional public accreditation of the programs. There is an opinion in the pedagogical community that the reports do not allow to create.

The necessity of the development of effective internal control systems of the quality of education is determined by the respond to external challenges as well as the strengthening of its position. In my opinion the story seems to be rather exhausted. There are many reports which often repeat the idea of the futility of such systems, as well as of the bureaucratizing creative activities. Nevertheless, I venture to suggest that the problem lies not in the systems, but in the nature of their use as well as in the initial relation to them. Why should one blame the tool, if the main reason for failures lies in the peculiarities of its application? One of the reasons of the paradoxical situation in which are the quality management systems of art education, but not the quality itself, lies in our opinion in the problem of the undeveloped quality management ideology of art education in centers of the supplementary education. The passion to formal procedures has not allowed paying attention to the philosophical and methodological aspects of the problem.

Meanwhile, the emphasis on the culture of the quality of the learning process in the institution, on the culture of relations within, for example, the choral studio of the Palace of the Children's (Youth)

Creativity, on the integrity of benchmarks of teachers-specialists in their field as a systemic factor in the quality of education are the current trends.

On the base of some common features of the management in social systems we can conclude that as the basis of the quality management of the education in the institution it is advisable to put the holistic motivational approach, which corresponds to the modern art education development priorities to a greater extent than any other. This approach assumes that on the one hand there is a common system of values; and on the other hand there are the motivators which support conscious acceptance of these values.

The neglect of one of the components restricts the management capabilities. Thus the absence of common values for the subjects of the educational process leads to their autonomy, disunity, as well as to the dominance of administrative approaches in the management. "In contrast to the motive, which being always mine, yours or his one, isolates the individual, living world, the value attaches on the contrary an individual to some supra-individual community and integrity, but ... does not dissolve it in this community, but, paradoxically, individualizes it"[1].

Supra-individuality of the values allows the creating of an educational space in the institution, as well as the uniting of all the group teachers in searching of some effective ways to achieve the goals of the art education. However, despite the attractiveness of the control based on the absence of actual values, the active motivators lead to declarativeness of the value systems. According to the analysis, the important guidelines of modern values are clearly identified, these are:
- the formation of self-image of the educational institution, providing the system integration of education and science, continuous improvement of the quality of educational activities;
- qualification and motivation of the group teachers;
- continuous qualitative improvement of the educational methodical and logistical basis of the educational process;
- implementation of modern learning technologies.

The justice of this choice cannot be denied, however, all these guidelines often remain only slogans and haven't any impact on the real reinforcement of motivation.

The value and motivational approach allows the changing of the situation in which the potential of key participants of the artistic process (i.e. group members, teachers) is leveled in practice as a guarantee of the quality of art education. The basis of this approach is presented by some rules, which implementation allows improving the management of the quality of education in the institution, changing of the overall atmosphere as well as the attitude to changes and their role in them.

The first principle of the value and motivational approach consists in the polysubjectivity of the quality management. One of the principles of the management states that the leadership is a trigger mechanism of the quality system. Thus the administrative office (i.e. the director and his/her team) acts as the leading guarantors of quality. However ideally the leadership provides the creation of the conditions under which it is possible and necessary to demonstrate the initiative of subjects of the artistic process (i.e. the teaching staff, group members).

Implementation of new standards requires the establishment of elite educational institutions, where professionals are trained to work in an innovative environment being capable to use the integration technologies and taking responsibility for the results of the artistic education. Unfortunately not all group teachers associate their labor efficiency with the quality of the educational work of the educational center and with the training of graduates in groups. Perhaps this problem is due to the quite low profile which is actually assigned to them in the quality assurance system. Nevertheless this situation must be liquidated in the modern environment.

For example in our time the choral studio of the Palace of Children (Youth) Creativity expands its communicative space and keeps the internal interdisciplinary dialogue open. The center of scientific and educational space can't be a separate department of the artistic direction or a separate group. It isn't the future but the present for successive interdisciplinary projects.

FIGURE 1. Concert of chorus "Melody" MUDOD DD (Yu)T of Perm at the International festival of arts devoted to Samantha Smith's (USA) visit to the USSR (1986)

The next principle is the development of the human capital. The implementation of this principle is often associated with the policy of recruitment of the most talented children and teachers. A more efficient mechanism consists in the investment in the training of the group teachers, in the constant sharpening of their skills and knowledge, in the development of the educational ways options as well as in the development of an effective system of training. One mustn't forget that investment in professionals, i.e. in their professional and personal development, improves the quality of human resources. A new by nature artistic process can be organized and evaluated only by a qualified teacher who constantly sustains his/her efforts related to his/her professional search and self-improvement.

The new look of the supplementary education of the younger generation requires undoubtedly the appropriate level of the pedagogical qualification. The system which provides that level can't be formal at the same time. The artistic programs which are mastered by the group teachers for the improvement of their qualification, must be competent and be based on a modular principle and network forms of organization of the educational process, they must allow to shape the individual curriculum and to solve the problem of every teacher.

Another principle of the value and motivational approach consists in the integration of science and education of the younger generation. Within the frameworks of this integration it is possible to build a coherent educational and professional environment which could recreate the artistic activity and promote personal and professional self-determination of the group members, as well as the development of innovation infrastructure for the improvement of the quality of research and development on the part of the teaching staff.

The openness is the next principle. The modern supplementary education becomes an open, socially oriented system. First and foremost the principle of openness provides the access to the information about all activities of the training centers. At the second the openness enables the integration of a Palace of Creativity in the educational space related to the implementation of the designated mission.

The next principle of the value and motivational approach consists in the improving of the quality of the content and education technology. The rate of obsolescence and update is associated with the constant changes in the artistic program and the organization of the educational process. The modern training can be high-qualified only through the continuous updating of its content and improving of the educational technology.

Nowadays there is a demand of the "model of knowledge as a conversation in the community". The teachers and the younger generation are considered to be the partners. At the core of the work lies the team activity, which is aimed to solve the creative problems and which hasn't any ready answers for the students. Such are the modern institutions. The communication between the generations becomes dialogical; it is constructed as a collaborative searching for the truth.

Motivation of the subjects of the artistic process (i.e. group members, teaching staff) consists in improving of the quality of the performance (certificates, diplomas, awards, titles) activities.

This rule includes an emphasis on the determinants of the training to obtain the desired results. When considering this principle as a system, the capacity of the teachers in the institution turns into a basic resource of improving of the quality of artistic education. Unfortunately, the system of motivational reinforcement of pedagogical efforts and achievements in educational institutions is underdeveloped in our time. The raise in salaries is a necessary protective measure, but it looks "naive" in terms of the motivational dynamics of a professional as a person (the nature of needs is not mechanistic). In particular one should not only create conditions for the varied salary and separation of the professional roles of teachers' circles, but also should promote the appearance of innovative "plots" of the application of their activity.

The basis of the quality management of artistic education in the municipal institution of supplementary education of the younger generation is a value-motivational approach; and thus the ensuring of the professional development of the subjects of the artistic process (i.e. the teaching staff, the group members) is the primary goal of the quality management systems. We have developed the theme of the research on the example of the principles.

References

1. Vasilyuk F.E. Psychology of experiences. Analysis of overcoming the critical situations. M.: MGU, 1984. - p. 13.
2. Shvydkoi M.E. From what class? // Musical Review. - №4 (April) (381). - 2015. - p. 2.

The Issues of Formation of the Graphic Literacy among the Teachers of Fine Art and Engineering Graphicss

Erkin Ruziev, Rustam Latipov
Urgench State University, Urgench, Uzbekistan

The modern society lives in the era when the unseen rate of growth of information flow is observed both in economic and social spheres. The use of the information flow and the wide reduction to practice of its contemporary transmission means began the evolution of the period of the informatization.

Nowadays the human life is full of different types of information such as sound, text or video. The graphic information is also one of these types of information. Many people understand and accept the graphic information in the form of images carved in stones, and lately in paper, canvas, marble and other materials which express the real world through pictures, photos, charts and drafts [1]. The others imagine it in form of schemes, drafts, charts, images on chart (figurative image) or shape. [2].

The images on various carriers of information such as paper, film, tracing paper, cardboard, canvas, hardboard, glass, wall etc. are also considered to be the graphic information [3].

Like every object the graphic information has its own properties. From the viewpoint of teachers of Fine Art and Engineering Graphics for elementary and high schools as well as the vocational colleges, there are the following properties of the graphic information: objectivity, reliability, completeness, clarity, simplicity, brevity etc.

1. *Objectivity of graphic information.* Objectivity is the free of one's opinion existence which doesn't depend on anyone. Information is here the reflection of the external objective world. Information is considered to be objective if it is free of one's opinion, subjectivity and thought. We can get the objective graphic information with the help of measuring instruments.
2. *Reliability of graphic information.* The graphic information is considered to be reliable if it reflects the real gauge (parameters) of something. The reliable information helps to perform the assigned tasks suitably. The lack of the reliability of the graphic information can be reflected through the following causes:
   – Changing of the subjective properties intentionally (misinformation) or unintentionally;
   – As a result of the insufficient accurateness of the registered image.
3. *Completeness of graphic information.* The graphic information is considered to be complete if it is sufficient for understanding and performance of a solution on its base. The incompleteness of the graphic information can lead to a wrong solution or drawing.

4. *Precision of graphic information* is identified by the degree of similarity of a draft (detail) to its original state.

With the informatization of the society and development of the graphic information, the formation of the graphic literacy among the future specialists is considered to be one of the tasks of higher vocational and professional education system.

The analysis of the situation concerning the training of the pedagogical human resource in higher educational institutions showed the possibility of training of highly competitive human resource of thorough knowledge on the base of the graphics culture which is formed in Schools (Academic lyceums, Professional Colleges) and has enough high quality. But for some reasons it is being observed that the degree of the graphic knowledge of the General Educational Elementary and High school graduates is decreasing. Thus, as a result it leads to the necessity of starting of the training from the very first steps in order to form the graphic literacy of students at universities.

In the recent years, the time intended for the studying of graphic courses (drawing geometry, drawing, computer graphics, and training method of the engineering graphics) in the educational plans and regulations for preparing the bachelors with a degree in "Fine Art and Engineering Graphics" have been dramatically decreased without any reason [5]. Considering that the drawing geometry and drawing courses are taught to the students within the initial semesters, the main part of the intended time is being spent not on the getting of some fundamental knowledge and skills, but only on the adaptation of the students' mind to the new requirements of the higher educational system. On the other side, the modern education requires from the students a high degree of preparation and profound knowledge on graphics. The quality of this preparation is accomplished through the graphics course cycles which develop the spatial perception and concepts of students, talents and constructing creativities of specialists, as well as their professional and graphic culture. Each of these graphic courses serves for the formation the correct concepts on the studied object being, for development of the theoretical and practical skills which are necessary for reading of drafts and other technical documents and suitable implementation of them.

It is impossible to learn the complex graphics and form the necessary skills without any proper and profound knowledge and education. The graphic literacy is understood as a level of knowledge. Quality of the literacy of a future specialist is graded through knowledge and the formed personal qualifications directed towards the performance of social and vocational tasks [6].

The grading method of the graphic literacy of a future specialist can be done through one of the following two ways:
- on the base of the graphic literacy of a graduated student through the concluding results of educational processes in higher educational institutions, and
- the graphic cultural knowledge of a future specialist which are acquired by the social information, his/her entrepreneurship, creative abilities as well as the teamwork abilities which show his/her ability to get along well with the other team members and make the connections within the team proving his/her social skills as a team player for performing a teamwork.

The "Graphic culture" is a wider and more diversified concept than the "Graphic literacy". In its wide meaning it can be understood as the summary of the accomplishments and achievements of the manhood in processing and transporting pf the information in a graphic view [7].

Having explored the graphic preparation as a result of the multistage complex process which has the various degrees of the development process of the formation of the graphic culture (from the learning of simple graphic elements till the studying them thoroughly with the purpose of practicing in professional and creative activities), M.V. Lagunova [8] shows the following degrees of the graphic culture in the education:
- Elementary graphic literacy;
- Functional graphic literacy;
- Graphic literacy;
- Possessing of profound professional knowledge;
- Graphics culture.

Aaccording to M.V. Lagunova, the elementary graphic literacy means the awareness by a student of the elementary laws of drawing theory and ways of the identifying and defining them on the base of the general geometric knowledge, practical skills of performing of drawing and skills of using the drawing devices on an advanced level and in practical matters.

The graphics culture as the main component of the General culture includes the following: a high degree and deep knowledge in the sphere of the isualization (grading with eyes approximately, measuring), skills and qualifications; understanding of the mechanism of using of the graphic images effectively concerning the professional problems and solutions; reflection and commenting of the results in an appropriate esthetic view and form.

The creative activity in the preparation for the professional activity of a future specialist, as well as the continuous independent studying and increasing of the level of knowledge in the sphere of the graphic information technology sets the degree of graphic culture.

So, in order to train the teachers of the Fine Art and Engineering graphic who have a high and deep graphic culture as one advantage along with others, first of all, we need to get to the point in which we can train the graphic literacy of the students in a sufficient degree. As we noted above, the graphic literacy is a generalized term and it is not limited only to the drawing field, but also spread to the field which shows the sufficient engineering (construction engineering, construction architecture etc.), designing and computer graphics knowledge of a specialist at a high level.

The latest changes in the standards of the training of the bachelors of the specialization "Fine Art and Engineering Graphics", in particular the sample education plans, show a dramatic decrease of the quantity of time intended for the graphic course cycles and semesters for learning. The engineering graphics courses also are decreased without any reason. In particular it concerns the Drawing course; the new educational plan has only 230 academic hours intended to the training within three semesters (the 4th- the 6th). It is necessary that teachers trained especially for the Vocational colleges are appropriate prepared and have profound knowledge on usual drawing course along with the knowledge on the automotive engineering course, construction, topographic drafts and schemes course up to the proper mark in order to be ready to teach these courses and to train the future specialists in this field. Although if we work according to the new offered regulation which states that drawing course should be taught only during 3 semesters we, will be able to teach and introduce only basic and general concepts. The specialists will meet the requirements and standards of the modern education and society only if we teach the drawing course for 8 semesters (starting from the 1st) including the courses of the automotive production and construction, as well as if the students perform their term.

New standards state that the bachelors of the specialty "Fine art and engineering graphics" will teach the graphic course cycles and in particular the different directions of the fine art course not only in the elementary schools and high schools but also in vocational and professional colleges. The perspective course plays an important role in training of teachers of fine art. Students draw and paint according to the laws and rules of perspective starting from the 1st semester. (The perspective course itself is planned to start in the 3rd semester). In the current educational plans and regulations, the perspective course is taught within the Drawing Geometry course, and even worse, as a last chapter of it and only during the 3rd semester. This situation has a disastrous influence on getting of knowledge on drawing geometry in order to study the art perspective course. The art perspective course in training of painters differs from the traditional art perspective courses with the amount of materials which should be studied additionally. Therefore, we think that it is compulsory to study this course starting from the 1st semester under the name of the "Practical Art Perspective".

From the 2nd half of the 20th century, design started to play a distinguished role in the different spheres of activity. Responding to this need for specialists, the design and its different directions training were introduced in the programs of educational institutions. In our country along with that it would be vital to train the different directions of design in the vocational and professional colleges according to their specialties: graphic design, fashion design, industrial design and books design. We believe that it is not necessary to accent and point out that these courses in the vocational/professional colleges will have the possibility of leading to graduates of the programs of the specialty "Fine Art and Engineering Graphics". The reason for it is the fact that any good designer is expected to be a good painter and to have the construction skills. The best of the pedagogical staff and human resource and those who learned the materials of the courses concerning the specialty and have profound knowledge and skills are considered to be the bachelors of the program of the specialty "Fine Art and Engineering Graphics". Therefore, the introduction of the design course into the structure of the educational module of the program will also help to raise the quality of the design courses in the vocational/professional colleges.

Graphics courses are considered to be one of the most important elements which play the main role in the formation of the professional and graphic culture of the students of the Bachelor program of the specialty the "Fine Art and Engineering Graphics". Therefore, on the base of the modern standards and requirements for education, it is necessary to raise the effectiveness of training through activities both in- and out-class-activities, using the innovational pedagogical methods and modern computer technologies.

The using of the computer technologies during the lectures allows us to introduce a big volume of information about the graphic objects in a short time, including the clear presentation of their spatial shapes and disclosing the dynamics of the surface emersion thanks to the multimedia elements. It develops the spatial perception of students and ability to sip the graphic information from the screens. So, the use of the lectures in the form of presentation assists in the formation of the graphic culture of the students.

During the practical lessons of graphics courses the main attention is usually paid to the solving of some graphic problems as well as to the consolidating the theoretical materials which had been got during the lectures. During the courses of Drawing Geometry the students study to compare the spatial objects with their linear image, i.e. the projections. The projection method lies at the root of any drawing of the automobile production, architectural construction and topographic drafts. The solving of positional and metric problems within the Drawing Geometry helps not only to develop the spatial perception of the students, but also promote the gaining of skills on the algorithmic and logical approach in the solving of problems concerning some real collocation and measures of objects. In order to raise the educational effectiveness during the practical lessons and trainings it is recommended to use the work-books with the in advance prepared graphic tasks. Thus, it will help the students to gain the practical skills on constructing of different graphic images and drafts during the practical lessons, to find out the convenient ways of solving of the graphic problems, as well as to form their graphic culture.

In conclusion we can say that it is necessary to review the concept of the educational standards concerning the bachelor-level program of the specialty "Fine Art and Engineering Graphics", as well as concerning the supply of study and method with materials and mediums, the inclusion of use of the informational and computer technology, the opportunities of putting in practice of the computer graphic resources, the structure module of the course on the base of the educational needs and requirements, and time intended for different courses.

References

1. Types of information and their properties. [Electronic resource] //http://ru.wikibooks.org
2. Pictorial (graphical) information. [Electronic resource] //http://www.gisa.ru.13223.html
3. Peculiarities of the graphical information and methods of coding. [Electronic resource] //http://tid.com.ua/tid1/addonres.php?id=3217
4. Jikhareva A.N. Graphical information and processing. [Electronic resource]/ N. A. Jikhareva //http://revolution.fllbest.ru/programming/00193778_0.html
5. Ruziev E.I. Role of the graphics coursetraining of teachers of the "Fine Art and Engineering Graphics". Training method of the Fine and Practical art, drawing teacher staff, T.: 2011
6. Ruziev E.I. Requirements for a drawing teacher. The problems of musical education, fine art and practical art course in the higher education system. Bukhara, 2010
7. Lomov B.F. The problems of the general, pedagogic and engineering psychology. M.: Pedagogika, 1991. — 296 pages.
8. Lagunova M.V. Graphic culture of engineer (Theory basis). Nizhny Novgorod, VGIPI, 2001. — 251 pages.

Law

Administrative Legal Relationships and Administrative Regulations in the Russian Legislation as Components of Administrative Treatment Mechanism Concerning Illegal Use of Drugs

Hui Wang

Saint-Petersburg State University of Aerospace Instrumentation, Saint-Petersburg, Russia

Abstract: *Legal norms form the principle foundations of a legal treatment mechanism. Administrative legal norms regulate administrative relationships both between federal executive authorities and authorities of federal subjects, local authorities and public institutions, NGOs, civil society organizations and other agencies interacting in conducting of the state anti-narcotics policy. [1]*

Keywords: *administrative regulations, administrative relationship, administrative treatment, illegal use of drugs.*

Administrative norms constituting the legal mechanism of treatment of illegal drug trafficking and use of narcotics regulate legal relationships in the following spheres:

1) activities of executive authorities at federal and regional levels, activities of local authorities (laying down the rules for state regulation of narcotics usage, regulating authorities and responsibilities of respective federal subjects, arranging interdepartmental cooperation in this field, etc.);

2) drug-addicts resocialization and social rehabilitation activities of nonprofit and medical organizations (public financing, regulation of activities, etc.)

3) imposing coercive measures on drug-addicted persons (i.e. bringing administrative actions). [2]

Legal theory abounds in varied systems of administrative regulations classification. For example, they are divided into substantive and procedural norms, protective and regulatory ones.

Substantive norms set forth rights, liabilities and duties of administrative relationships subjects and may be as well divided into prohibitive, binding, enabling, restrictive, incentive, registration, advisory and permitting regulations, which is conditioned by a norm's legal content. All the aforementioned types of regulations are applicable in the sphere this article dwells upon.

Procedural administrative norms regulate the procedure, the common rules, jurisdictional, law-enforcement and legislation activities of state governing authorities as well as other subjects of administrative law. Their content is covered at its fullest within the framework of administrative offenses proceedings arising from art. 6.9 of the Administrative Offenses Code of the Russian Federation. [3]

Protective norms establish obligatory rules of conduct, restrictions and prohibitions concerning specific actions. Regulatory norms involve such legal instruments as license and registration proceedings, notice of activities, etc.

This classification is fully applicable to rules of administrative law as one of the elements of the administrative legal treatment system of offenses concerning illegal drug trafficking and use of drugs. Prohibited use of narcotic substances without medical prescription contained in art. 40 of the Federal Law #3 dd. 01.08.1998 may serve as an example. [4]

It seems expedient to use the classification of both international and national regulatory acts based on the content of legal norms they set forth when studying specific aspects of anti-drug administrative law.

1) Documents establishing the basic principles of state policy in this sphere. These include international conventions, the Constitution on the Russian Federation, the Federal Law #3 dd. 01.08.1998, etc.

2) Regulatory legal acts defining liability for breach of regulations (the Criminal Code of the Russian Federation, the Administrative Offenses Code of the Russian Federation).

3) Documents setting forth general provisions, rights and duties of subjects of law (i.e. the Decree of the President of Russia #976 dd. 07.28.2004 concerning the Federal Drug Control Service).

4) Program and goal-setting legal acts aimed at improving the illegal drug use treatment mechanism (i.e. the Order of the Russian Government #299 dd. 04.15.2014, which approves the state anti-drug program of Russia, and the Resolution of the Russian Government #202-r dd. 02.14.2012, which confirms the action plan on improving the system of medical aid and rehabilitation of drug-dependent people and prescribes the establishment of the public prevention system in this sphere).

5) Legal acts regulating specific aspects of interaction between executive bodies, organizations (social and medical) acting withing the framework of the anti-drug policy (i.e. the Federal Law #7-F "On Non-profit Organizations" dd. 01.12.1996, the Law #323-FZ dd. 11.21.2011).

6) Branch-wise documents regulating specific aspects of anti-narcotic legal treatment mechanism (i.e. the Orders of the Russian Ministry of Health #327 dd. 08.23.1999 and #704 dd. 09.12.1988).

This division is rather relative as long as some documents may be sorted into more than one category.

Being the ultimate foundation of legislation, the Russian Constitution contains statutory provisions that play the key role in the country's legal system. As for anti-drug efforts, the Constitution lays down legal guidelines, separates the authorities of federal and regional state institutions, determines the substance of their activities. [5]

International regulatory acts on fighting drug trafficking generally contain indirect references to the need for a treatment system regarding this issue.

The Federal Law #3 dd. 01.08.1998 lays down the legal basis for the state anti-drug policy and regulates the main directions and the action strategy of the country in this sphere. Based on the analysis of international as well as Russia's anti-drug efforts this legislative act suggests the most optimal to date solution for problems in this field, a system of control over the respective activities and measures of legal treatment affecting both drug supply and demand.

However, a number of specific provisions laid down by the Federal Law #3 dd. 01.08.1998 could not be fulfilled without the conditions and the proceedings established by the Administrative Offenses Code of the Russian Federation.

The Administrative Offenses Code contains administrative legal regulations concerning the mechanism of illegal use of drugs treatment. These regulations determine the proceedings and general rules for law-enforcement activity carried out by state institutions and other subjects of law involved in conducting anti-drug policy.

As for the norms of procedure contained in the Administrative Offenses Code of the Russian Federation pertaining to illegal use of drugs, it is worth pointing out the following peculiarities and specifics of penalty infliction:

— drug control agencies are authorized to consider administrative offenses cases listed in sec. 2 art. 20.20 of the Administrative Code. Other cases are heard in court;

— officials of drug control agencies are authorized to draw up reports of administrative offenses listed in sec. 3 art. 20.20, sec. 3 art. 19.3, art. 6.9, art.20.22 of the Administrative Code;

— conducting of administrative investigation of such cases;

— penalty in form of up to 30 days under administrative arrest, etc.

Program and goal-setting legal acts relating to illegal use of narcotic substances are represented by programs, concepts and strategies. They determine the main improvement tracks of administrative legal treatment mechanism related to illegal use and trafficking of drugs, contain specific methods and forms of treatment, etc.

For instance, the Russian Concept of National Security approved by the President encompasses tasks related to prevention of illegal drug-trafficking, rehabilitation and medical treatment of drug-dependent citizens alongside such high-profile goals as securing life, health, the rights and the liberties enshrined in the Constitution. [6]

It is worth noting that the administrative relationships emerging in process of administrative legal treatment of drug use and trafficking have all the features inherent in administrative relationships in general. A distinct structure presents itself consisting of subjects and objects of legal relationships, normative content, hierarchy and authority, a state governing body as a mandatory subject, etc. Besides, these relationships have a wide variety of forms, such as protective and regulatory, coordinating and subordinate ones, etc.

Proceeding from theoretical conceptions concerning distinctive features, the structure and the substance of administrative legal relationships it is possible to point out the peculiarities of legal relationships in context of administrative legal treatment of illegal drug use and trafficking within the Russian legal framework:

— a significant number of subjects in legal relationships requires a system of interdepartmental cooperation;

— presence of the State Anti-Drug Committee – a subject in legal relationships possessing peculiar features and not included into the executive power structure but nevertheless exercising authority in this sphere;

— enabling norms that regulate the duties of the Federal Drug Control Service as the coordinator of ministries', departments' and organizations' activities related to resocialization and rehabilitation (with the exception of medical) of drug-addicted persons form the foundations for harmonized administrative legal relationships. Such relationships necessitate coordinated activities of several subjects of law in order to cut the demand for narcotics;

— increasing role of administrative legal relationship between executive bodies (local administrations) and NGOs dealing with social rehabilitation of drug users aimed at enhancing their support and controlling their activities;

— the result of a wrongdoing is not limited to protective legal relationships in form of bringing the offender to administrative responsibility, but also extends to legal inducement of a drug user to rehabilitation, medical treatment, etc.

The typical examples of regulatory administrative legal relations arising from legal treatment of illegal drug use and trafficking are as follows:

— legal relations within the Federal Drug Control Service (a strict hierarchy of territorial subdivisions);

— fulfillment of the duties of the Federal Drug Control Service through coordination activity; [7]

— conditioned state support to organizations dealing with resocialization and rehabilitation of drug-addicted individuals;

— the licensing procedure for medical organizations to deliver care to drug users;

— the procedure for regular medical and prophylactic check-up, etc.

The format of the article does not give room for a comprehensive sound analysis of this complex topic dealing with issues of administrative legal norms and legal relationships between subjects of law, which constitute an inseparable part of the mechanism of administrative legal treatment of illegal drug use. This topic has become a subject matter for many legal scholars and practicing lawyers, which can give an impetus to improving the legal framework in order to solve the most challenging social problems facing the Russian society.

References

1. Melehin, A. V. Theory of State and Law: a Manual/A. V. Melehin. – Moscow, 2007. – 640 pp.
2. Scientific practical paragraph-to-paragraph commentary to the Federal Law #323-FZ dd. 11.21.2011/Blagodir, A. L., Dubrovina I. L., Kirillovyh, A. A. et al.; ed. by Kirillovyh, A. A. – Moscow: Delovoi Dvor, 2012. – 600 pp.
3. The Administrative Offenses Code of the Russian Federation dd. December 30th 2001, #195-FZ// Legislation Bulletin of the Russian Federation, January 7th 2002, N 1 (part 1) art.1
4. The Federal Law #3-FZ dd. January 8th 1998 "On Narcotic and Psychotropic Substances" (as amended on July 25th 2002, January 10th and June 30th 2003, December 1st 2004, May 9th 2005, October 16th and 25th 2006)/
5. The Constitution of the Russian Federation Adopted by Referendum on December 12th 1993 // Rossiyskaya Gazeta, #237. December 25th 1993.
6. The Decree of the President of Russia #537 dd. 05.12.2009 "On the Concept of National Security of the Russian Federation to 2020"//Legislation Bulletin of the Russian Federation, – 2009, #20, art. 2444.
7. The Decree of the President of Russia #976 dd. 07.28.2004 " The Issues Concerning the Russian Federal Drug Control Service"//Legislation Bulletin of the Russian Federation, – 2004, #31, art. 3234.

Citizenship in International Law: Concept and Legal Content

Ksenia Tyurenkova[1], Karine Muravieva[2]
[1]Astrakhan State University, Astrakhan, Russia
[2]International Law Institute, Astrakhan, Russia

Abstract. The article deals with the definition of the essence and importance of citizenship. The author considers different scientific concepts concerning the definitions of the citizenship notion as well as the peculiarities of the citizenship, representing one of the types of relations governed by law – legal state of a person in citizenship.

Keywords: state, citizenship, nationality, foreigner.

1. The first part.

Today, citizenship is a central point of the relationship between the state and the individual. Generally, questions of nationality are regulated by domestic law, but now especially increased the role of international law. This is due to the increase in disputes between states over the legal status of specific individuals or specific groups. Therefore, in addition to the values in the inner sphere, it has an international aspect, since the acts of interstate cooperation. International law affects the whole group of issues of citizenship: its contents (the rights and duties of the individual and the state) institutions that are directly related to citizenship in the field of inter-state relations (diplomatic protection issue), acquisition and termination of citizenship. The last decade of the XX century is connected with the rapid development of the institution of citizenship. In the field of citizenship in international law there are certain standardized approaches some kinds of "standards".

Thus, citizenship is one of the most important human rights, the foundation of the legal status of the individual, not only within any State, but also in international communication. Only on the basis of citizenship of a certain state an individual can take advantage of the maximum level of the rights guaranteed by this citizenship.

So, with this in mind, the study of issues of citizenship has both theoretical and practical interest.

The term "citizenship" is derived from the word "city", "citizen." From the Latin «civis» was established the notion «civilis» - a civilian, state, which gave the origin of the notion "civilization". [2]

Such notions as "citizenship" and "citizen" have a long history. The first mention of certain categories of citizens living in different states, belong to the history of the Ancient East. For example, in the texts of the laws of the Babylonian king Hammurabi (1792-1750 BC.), There are some certain notions which are comparable with current categories in the area of citizenship. [1]

Antique citizenship was an alliance of free and equal citizens, based on collective ownership and exploitation of slaves. In ancient Greece, the term "citizen" refers to the legal status of a free man, the presence of his rights and freedoms. Greek citizenship as opposed to legal status of foreigners was determined primarily by the presence of political rights. Understanding by the ancient Greeks as the universal values of citizenship (for free citizens) as a right which cannot be alienated as a result of immigration in another country because of political and other internal strife is very interesting.

The idea of the special rights and duties of a citizen was highly developed in ancient Rome. Citizen was originally a resident of Rome, freeholder, and then, in the days of the Empire, Roman citizenship was extended to the first inhabitants of the Italian peninsula, and then on all other subjects of the Roman Empire. Roman had a civic duty and the right: in the army, took part in public meetings, in political life; it cannot be, for example, the crucified.

In feudal society, the notion "citizen" is a synonym of a free man, having a certain set of political and economic rights, giving way to the notion of "citizenship".

Nationality and citizenship are different in nature of the relationship of the individual and the state. If citizenship is characterized by one-way communication that expresses the responsibilities of the individual in relation to the monarch, the citizenship is characterized by two-way communication between the individual and the state, mutual rights, duties and responsibilities.

The concept of civil society was developed by the major thinkers of XVII-XIX centuries - John Locke, Jean-Jacques Rousseau, Kant. In the "social contract" Rousseau regarded citizenship as an expression of a certain nationality community persons in the state, giving them the right to participate in government. He believed that the citizen is not "having a share in the power of the sovereign," opposed to the citizens of a "how to obey the laws of the state" (the relationship between the terms "subject" and "citizen"). The thinkers of that time also linked the notion of "citizen" with the principle of private property. For example, Diderot believed that "property creates a citizen" and "Only the owner is a true citizen." The ideas of Rousseau and other supporters of the concept of natural law school influenced the "Declaration of the Rights of Man and of the Citizen", adopted in 1789 by the French National Assembly. This declaration officially introduced the notion of "citizen", which is associated with an idea of a legally free person, has the right to participate in political life.

Currently, the countries with the republican form of government usually uses the term "citizenship" in countries with monarchical form of government still uses the term "citizenship" [4].

A number of modern monarchies (Spain, Belgium, the Netherlands), the term "citizenship" in the constitutions and legislation substituted for the term "citizenship". Although some states with monarchal form of government (the UK, Sweden, Norway, Japan) have retained the term "citizenship", but it is the same in meaning to the term "citizenship". We can say that now citizenship is only a synonym with outdated nationality [8], with the same set of rights and freedoms, that is actually tantamount to citizenship. [7]

2. The second part.

Let us analyze the various points of view of national researchers about their understanding of the notion of "citizenship".

L. Oppenheim realized citizenship of an individual as the property of any person be subject, and because of this - a citizen of a particular state. [6]

V.M. Gassin considered citizenship as a set of public rights and obligations of the presence of which is due to his nationality. In his opinion, nationality - is an individual and the state, private submission of its law-making power, the legal fact, the state, the property of the person, similar to sex and age and is the basis of legal capacity in general. A subject is always, a citizen, regardless of its name in the domestic law of the State. [3]

Këssler called citizenship as "status of an individual and the state." Weiss wrote that "citizenship in the sense of international law is the technical term meaning affiliation of individuals, called citizens to a single state - the state of citizenship - as a member of the State, relation which gives State of the international rights and obligations of states in relation to other states," [11] .

From these statements it is clear that most scientists realized citizenship as belonging the individual to the state.

It should be noted that a number of modern Russian authors believe that citizenship - a stable legal relationship between a man and the state, expressed in the totality of their mutual rights and duties specified in law. [5]

Based on this definition of citizenship V.V. Maklakov and B.A. Strashun state that the citizen is under the sovereignty of the state, and the latter may require him to perform duties even if he resides abroad. The government, for its part, must protect the citizens on its territory and provide them protection when they are outside. [9]

However, some authors emphasize not only the legal but also the political nature of this relationship. So, G.I. Tunkin said that "...under the citizenship mean political and legal connection with the state, which is characterized by resistance to possess rights and duties of the state in relation to a person" [10]. A similar opinion is shared by several other authors [12].

In accordance with the Federal Law "On Citizenship of the Russian Federation" citizenship is defined as a stable legal connection with the state, expressed in the totality of their mutual rights, obligations and responsibilities, based on the recognition of and respect for the dignity, fundamental rights and freedoms.

It is worth noting that, now a day, foreigners have access to a certain extent the different rights enjoying by citizens of the State, but it is usually not available to foreigners to use political rights. Only citizens of their state can actively participate in the political life of the state.

Thus, from a theoretical point of view, the compromise and the most common definition of citizenship could be next. Citizenship is a special stable and continuous political and legal ties (accessory) the person to the state, legally expressed in the regulatory consolidation bases introduction, acquisition and loss of nationality, if there is a person gets the opportunity to enjoy all statutory rights and freedoms, as well as acquires obligations constituting together the legal status of a citizen of the state.

Citizenship is a legal reflection of the social relations established between the individual and the state. In the social understanding citizenship means assimilation of the person in a particular country, his awareness of his involvement in the affairs of society. The basis of nationality in favor actual social bond between man and society. Expression of the will of the state, this relationship takes on a new qualitative significance: it becomes a stable political and legal communication. At the same time the legal basis of nationality should reflect its social character. Legal citizenship properties are its resistance in time and space efficiency, the availability of formal proof of citizenship of the volume of rights and responsibilities distinguish citizens from aliens. Data grounds of nationality is characterized not only in national but also in international law.

References

1. Belkin A.A. Some doctrinal matters of citizenship // Jurisprudence. 2013. № 6.
2. Batting I.V. Philosophy. - Rostov-on-Don: Phoenix, 2008.
3. Hesse V.M. Citizenship, its establishment and termination. V.1. - St. Petersburg: True. 1959.
4. The course of international law. Textbook / Ed. Ushakov NA V.3. - M.: Science, 2010.
5. International law. Textbook / Ed. Ed. Kolosov Yu.M., Kuznetsov VI - M.: Lawyer. 2009.
6. Oppenheim, L. International Law. V.1, Semivolume 2. - M.: Foreign Literature 1949.
7. Russian Thesaurus / ed. Cheshko LA ed. 5th. - M.: Russian language, 1995.
8. Smirnova E.S. Europe: Evolution of Multiple views on the second half of the XX century. // Law and Politics, 2010. №1.
9. Strashun B.A., Maklakov V.V. et al. Constitutional law of foreign countries. Tom1-2. A common part. - M.: Beck, 2006.
10. Tunkin G.I. Citizenship of the USSR. - M.: Nauka, 1989.
11. Chernichenko S.V. International legal questions of citizenship. - M.: International Relations. 2011.
12. Shestakov L.N. et al. International law. - M.: Legal Literature. 2009.

Other Social Sciences

Comparative Analysis of Parameters of the Functional Readiness at Students of the Meliorative College

Galina Khasanova, Regina Niyazova
Uzbek State Institute of Physical Culture, Tashkent, Uzbekistan

Abstract. *The article deals with the problem of the improving of the professional-applied physical training (PAFT) of college students due to the ever increasing requirements for the education and subsequent employment of graduates of specialized educational institutions, as well as due to the requirements to the employees on the part of the agricultural sector, which dictates a necessity of raise of the level of some special motor skills with regard to their future careers.*

Keywords: *professional-applied physical training, morphological and functional parameters, sensitive periods, hemodynamics, heart rate (HR), blood pressure.*

Introduction. According to one of the leading theorists of the doctrine of the physical development of a person V. V. Bunak, the physical development is a complex of the functional properties which define a stock of physical forces of an organism. The physical development reflects the process of growth and development of an organism at separate stages of the post-natal ontogenesis during the transformation of a genotype into a phenotype. It is well-known that the influence of the genetic program and factors of the environment on the physical development is unequal during various age periods. It is shown that the total sizes of a body depend on its length and weight, as well as the circle of thorax [1, 4, 5, 7].

The purpose of the study consists in the increase of the level of the morphological and functional capabilities through the rational allocation of the professional-applied physical training of students during the training sessions at a reclamation college.

The applied **methods** are the following: analysis of literature, morphological and functional measurements, pedagogical control tests, methods of the mathematical statistics.

As to the **organization of the research**, with a view to the practical implementation of this goal the study was organized and conducted several basic experiments in two secondary special and vocational educational institutions, involving 120 students.

Three educational groups of the Nukusky meliorative and water management college formed the experimental group (EG consisted of 60 persons), and another three groups consisted of the first-year students of the Tashkent meliorative and water management college formed the control group (CG consisted of 60 persons).

As it is noted by several authors [2, 3, 7] **the Results of the research** should be the methodological basis for simultaneous training of motor skills of students of an institution or an athlete at all stages of the long-term training, the priority development of the age-appropriate individual qualities.

The studied age range has a scope for the further improvement and development of just those qualities which could provide some pedagogical effect, and which are not designed to the fundamentally influence or modification of the age development patterns of some aspect of the motor function of students.

In the Table 1 it is visually shown that the investigated by us age period is an optimum for the development of the power opportunities and aerobic endurance (+) and it allows to continue the development of the coordination abilities and mobility of joints (±) of young men of 15-18 years.

TABLE 1. Typical sensitive periods of altered growth and weight parameters and development of the physical parameters of young men of 15-18 years old

№	Physical parameters	Age (years old)			
		15	16	17	18
1	Body length	+	±	±	±
2	Body weight	+	±	±	±
3	Maximum force	-	+	+	+
4	Quickness	+	+	+	+
5	Endurance (aerobic opportunities)	+	+	+	+
6	High-speed endurance	+	+	+	+
7	Anaerobic opportunities	+	+	+	+
8	Coordination abilities	±	±	±	±
9	Flexibility	±	±	±	±

The distinctions of the average sizes of the students' anthropometrical parameters which reflect forward process of the physical development (Table 2) were revealed by means of the selective and statistical method, as well as the method of the dynamic individual supervision.

The obtained data testify that the offered program of training contributed to the increase of the level of the physical development of students from the EG. The studied parameters of the students from the EG were not only evened to the 4th semester, but also exceeded a little in growth (177,1±5,47 and 176,9±5,19 cm) as for the fellow students from the CG at the final stage of training (the 6th semester); the students from the EG surpassed also authentically in body weight parameters (65,7±3,97 and 63,4±4,27, at P<0,05).

According to the experts [1, 5, 6] the important parameter of the hemodynamics, i.e. the heart rate (HR) and blood pressure, undergo the regular changes with the advancing age of the organism as well as the effects of physical exercises [2, 3, 7].

The pulse frequency decreases with advancing age, so according to V.F. Balashov [1], L.P. Volkov [2], O.A. Rihsieva et al. [5], D.A. Farber [6] the heart rate of boys of 15-16 years old who are not athletes is 76.2 ± 0.1, then the boys of 17-18 year old have already 72,0 ± 0.3 beats / min. As it is noted (Table 2) the heart rate of the 3rd year students from the EG is quite low (70,3 ± 1,47 beats / min) and significantly lower (P <0.05) than the parameters of the students from the CG (75,1 ± 2, 26). According to our observations a moderate bradycardia caused by the physical training prevents the aging of the myocardium and is of great curative importance.

The blood pressure is an important parameter of the cardiovascular system and is recognized in physiology and sports medicine as an integral indicator of the hemodynamics. A significant difference was observed (104/64 and 105/65, respectively) concerning the average maximum and minimum blood pressure in the EG and CG. Although the density of the results is more pronounced among the representatives of the EG (V=1,1% and V=3,3%, respectively). The results obtained by years of study, as well as the data which were gained by a number of physiologists [1,5,6] indicate the boys form the EG with the optimal cardio vascular parameters, regardless of the higher volume and intensity of the load which is carried out during the studying and training sessions intraday.

The state of the external respiration along with the physical development is essential for the assessment of the physical and functional readiness of an organism. According to the experts [1, 2, 3, 5, 6], the value of the vital capacity plays an important role in the mechanics of the external respiration and not only indicates the potential of human increase tidal volume, but also helps to estimate the required effort to provide ventilation.

The results of the analysis of our complex research suggest that all three measures, namely the vital capacity (4043,9 ± 105,6 and 3802,7 ± 113,3 respectively), chest expansion (9,43 ± 0.57 and 9,01 ± 0.89 respectively), and chest circumference (88,52 ± 1,33 and 86,7 ± 3,02 respectively) characterize the potential of the external respiration. The parameters of the students from the experimental group became significantly higher ($P < 0.05$) by the end of the sixth semester; this fact indicates a significant improvement of the motor and respiratory reflexes which are based on the proprioceptive afferent.

The improvement of the respiratory function of students of the EG contributed to the ordered structure of the distribution of the basic sports facilities during the training process and breakout sessions with the increased physical activity which is aimed at the development and improvement of the speed-strength and aerobic endurance.

TABLE 2 Dynamics of the morphological and functional parameters of the students from the experimental and control groups at the stages of their study (6th semester) (n = 140)

№	Parameter	The 1st year (15-16 years)		The 2nd year (15-16 years)		The 3rd year (15-16 years)	
		EG	CG	EG	CG	EG	CG
1.	Body weight (kg)	51,9±2,75 V=5,29%	50,4±2,97 V=5,89%	56,7±1,44 V=2,53%	53,2±3,12 V=5,86%	60,8±21,97 V=36,13%	57,4±3,27 V=5,6%
2.	Body length (cm)	169,1±3,33 V=1,96%	169,7±3,95 V=2,32%	172,9±2,98 V=1,72%	171,9±3,02 V=1,75	174,4±2,47 V=1,41%	173,1±4,19 V=2,42%
3.	Heart rate (Beats Per Minute)	74,3±2,07 V=4,05%	77,3±3,23 V=4,17%	72,1±1,04 V=1,44%	75,3±2,79 V=3,70%	70,3±1,47 V=2,09%	75,1±2,26 V=3%
4.	Systolic blood pressure (milliliters of mercury column)	103,37±1,97 V=1,9%	102,3±3,25 V=3,17%	104,33±1,48 V=1,41%	103,7±3,07 V=2,96%	104,23±1,47 V=1,41%	105,12±3,44 V=3,27%
	Diastolic pressure (milliliters of mercury column)	62,3±0,97 V=1,55%	63,5±1,16 V=1,82%	64,1±0,88 V=1,37%	65,7±1,97 V=2,99%	64,3±0,73 V=1,1%	65,2±1,55 V=2,3%

5.	Vital capacity (cubic milliliters)	3383±87,6 V= 0,02%	3090±103,3 V= 3,34%	3577±78,2 V= 2,18%	3292±137,1 V= 4,16%	3743±65,6 V= 1,7%	3402±113,3 V= 3,33%
6.	Chest circumference (cm)	84,76±2,08 V= 2,24%	80,35±3,17 V= 3,94%	86,43±1,91 V= 2,20%	82,1±2,89 V= 3,52%	88,12±1,33 V= 1,5%	84,7±3,02 V= 3,56%
7.	Chest expansion (cm)	8,69±0,87 V= 10,01%	8,39±1,07 V= 12,75%	9,03±0,565 V= 6,25%	8,55±0,98 V= 11,46%	9,23±0,57 V= 6,17%	8,61±0,89 V= 10,3%

Thus, quite a pronounced parallelism of students from the EG in changing of their parameters of height and weight as well as the parameters of the physical development such as hemodynamics, vital capacity, and chest circumference, indicates the beneficial effect of the proposed by us set of exercises for the professional-applied physical training of integrated control of the morphological and functional development of the organisms of students of the reclamation college.

References

1. Balashova V.F. Human Physiology. Moscow, - Physical Culture. 2007. - 375 p.
2. Volkov L.P. Theory and methodology for child and youth sport. Kiev. Olympic literature. 2002. - 294 p.
3. Koshbahtiev I.A. Fundamentals of the improving of the physical education of students. Tashkent, 1994. – 105 p.
4. Human Morphology. / Edited by B.A. Nikitiuk and V.P. Chtetsova. The 2-nd edition, revised and enlarged / Publisher of the Moscow University. M.: 1990. - 343 p.
5. Rihsieva A.A., Nasretdinov F.N., Rihsieva L.I. The physical condition of schoolchildren and youth sports. – Tashkent: Ibn Sina, 1992. – 152 p.
6. Physiology teen / Edited. YES. Farber. Moscow: Education, 1988. – 207 p.
7. Kholodov J.K. The theory and methodology of the physical education and sport: Textbook. - Moscow: Soviet Sport, 2008. - 480 p.

HUMANITIES

Arts

Glimpses of History of the Formation of the Orchestra of Turkmen National Musical Instruments

Bahar Goshayeva

Turkmen National Conservatory, Ashgabat, Turkmenistan

Abstract. *The musical culture of each nation is unique. For example, the balalaika is a purely Russian folk music instrument, the dutar is a Turkmenian one, etc. The repertoire of the orchestra includes not only works of local authors and the masterpieces of foreign classics, but also some works of authors from the other countries and nations. Thanks to the creation of new musical instruments it is possible to expand the repertoire. The composers created their works or adaptations of some folk tunes and songs for orchestra of folk instruments in a certain region taking into account the specificity, range and texture of each instrument. Our attention will be particularly focused on these methods and secrets. The research will be also focused on the history of the first orchestra of national instruments named after Purli Sariyev, the musical instruments which are part of it, the origin of new models of the reconstructed instruments, their advantages and disadvantages, and the people, without whom the foundation of the orchestra would be impossible, and who gave many years of his life to the improving the orchestra. The collected information will serve as the educational material for teachers, students and students of special schools.*

Keywords: *music, creativity, composer, orchestra, musician, musical national instruments, reconstruction.*

Grandiose changes took place in the country in all spheres of activity including the culture after the official proclamation of the Turkmen republic in 1924. Among other issues the problem of the indispensable cultural growth of the nation was also brought up within the 1st Constituent congress which was historical for the country. This issue concerned especially the countryside population. The attraction of well-known creative persons of the country was of great necessity for this purpose. The well-known and popular writers such as Kyormolla, Mollamurt, Durdy Gylych, Atakopek Mergen, Nury Annagylych, Ata Salyh, as well as the musicians and bakhshies such as Tachmammet Suhangulyev, Hally bakhshy, Sary bakhshy, Mylly Tachmyradov, Amangeldy Gonibekov, Purli Saryev, Oraz Salyr, Nobat bakhshy, Garly bakhshy, Sahy Dzhepbarov put their hands to the plow in a very responsible manner. The first Turkmen radio began its broadcasting in the capital. At the new drama theatre the ensemble of national musical instruments and bakhshies was found [1].

Due to the shortage of competent experts-musicians a decision was made to invite the composers and musicologists from the different union republics of the USSR to Turkmenia.

The graduate of the Petrograd music conservatory, musicologist and orientalist V.A. Uspensky was one of them. The scientific expedition under his guidance began the work in August of 1925 with the territory of the old Merv (nowadays Mary). During the period of 1925-1929 the expedition having studied all the territory of Turkmenia gathered a considerable quantity of the most valuable musical material which formed further the base for the co-written with the musicologist V.M. Beljaev the two-volume book "The Turkmen music" [3]. The impact of works of V.A. Uspensky on the further development of the Turkmen musical culture cannot be overstated. On the base of the gathered and deciphered by him musical material, such masterpieces were further created by B. Shehter, K. Korchmarev, A. Shaposhnikov, V. Muhatov and many other composers as the operas "Shasenem and Garib", "Zohre and Tahir" etc.

V.A.Uspensky with the guide

It can be said that the national professional musical culture which mastered achievements of progressive world art grew on the fertile soil of the national musical creativity in the Soviet Turkmenia. In republic operas, symphonic and chamber products, the songs created by Turkmen composers are executed. A big help in development of difficult professional forms and genres was rendered to republic by the teachers, composers, executors from the fraternal countries.

Thus the Art technical school which included the drama and musical departments was launched in Ashkhabad in 1929. Besides due to the insufficiency of professional staff such well-known musicians and composers as D. Ahsharumov, V. Tovstoluzhsky, V. Sardarjan, E. Konchevsky, A. Berger arrived here from different union republics of the USSR. As for the national music such khalipa (Turkmens - instructors), as Nobat bakhshy, Mylly Tachmyradov, Purli Saryev, and Garly bakhshy gave here their lessons. That time in the Art technical school such people as Ashyr Kuliev, Veli Ahmedov, Chary Tachmammedov, Shamyrad Gurbannepesov, and Sapar Mamiev were lucky to study; all of them are considered to be the luminaries nowadays.

In the thirties of the last century the creative unions were established in Turkmenia including the Union of composers of Turkmenistan, which was headed by the honored worker of culture R. A. Ivanov[1]. Although the membership of the organization at the beginning of its activity was not numerous, further it was coopted with the graduates of the Moscow and Tashkent music conservatories. The following competent musicians and composers returned to the country having graduated these higher education establishments: Veli Muhatov, Orazmuhammet Gurbannyjazov, Nury Muhatov, Chary Artykov, Durdy Nuryev, Matvej Ravich, Aman Agadzhikov, Nury Halmammedov, Redzhep Allajarov, Redzhep Redzhepov.

The musical-creative collectives of the republic began to appear in the end of the forties. So, the history of creation of the orchestra of national musical instruments began with the ensemble of musicians and bakhshies under the guidance of Sahy Dzhepbarov, which was found in 1937 and served at the State radio. The idea of creation of the first in Turkmenia experimental orchestra of Turkmen national musical instruments was coined by professor D. Ahsharumov and S. Tumanjan who worked that time with the symphonic band at the State musical school. Then it was decided that the collective would be headed by the teacher of the department

---

1 R. A. Ivanov is placed on one of the front ranks in the history of development and formation of the musical culture of the Soviet Turkmenistan. He was the first Russian highly educated musician and conductor, which pedagogical activity began in pre-revolutionary Turkmenistan.

The first cast of the orchestra of Turkmen national musical instruments.
The third man from the left is Purli Saryev

of the Turkmen national musical instruments and well-known musician Purli Saryev[1].

Geldy Ugurliev was called over to the cast by the teacher, and Setrak Tumanjan was called over by the main conductor, because by that time he had got a wide experience of work with the orchestra. S. Tumanjan, being also the expert on the manufacturing and restoration of stringed and bowed musical instruments, proposed to improve the gidzhak groups of the orchestra. So, thanks to it some changes took place: the traditional alto, bass as well as the contrabass were replaced by gidzhaks. Along with that S. Tumanjan planned the reconstruction of the dutar groups. Unfortunately, due to his death this intention was continued by his successors.

In 1940 the orchestra was transferred in the philharmonic society under the guidance of the Honored artist of Turkmenistan P. Saryev and the honored artist of Turkmenistan G.M. Arakeljan.

The continuation of works on the reconstruction of the instruments was necessary. Such experts as Geldy Ugurliev, artist Bjashim Nurali, Bagrat[2] were involved in this project. Unfortunately, the instruments they had created were imperfect and inappropriate to the requirements of the orchestra; the project required again for the search of new models and experts, who could develop those models. Thus A. I. Petrosyan and V. Didenko were invited from Tashkent as the experts. Having understood all the nuances of the Turkmen national musical instruments in a proper manner, they created dutars and gidzhaks which are being used in the orchestra up to now. These are the orchestral tonic, alto and bass dutars, group of plektor dutars[3] and a four-string gidzhak, while a traditional one has three strings. Thanks to these instruments the orchestra began to sound in a new manner and the repertoire was

---

[1] The honored artist of the TSSR. A well-known composer, melodist and teacher; went down in history of the Turkmen musical culture as a talented dutar player and gidzhak player.

[2] Experimental musical instruments of these experts had been kept up to now in the Museum of the national instruments of the Turkmen national conservatory.

[3] Version of tars. The sound on these dutars is taken through mediator.

extended due to these innovations; then the more serious and responsible tasks were possible to put [2].

The repertoire of the orchestra included both the adaptations of national melodies such as "Novai", "Nergiz", "Dilber", and the specially created for the orchestra of the Turkmen national musical instruments works, such as "Suite" and "Overture" by G. M. Arakeljan, march "Friends" by A.Kuliev and many others. Thus, thanks to the variety of its repertoire the orchestra had its listeners.

The first international on-scene performance took place in 1944. The orchestra takes part in the cultural decade of the people of the Central Asia in Tashkent. That experience of the performances inspired the collective to the further growth and perfection.

In 1948 a young composer and conductor Kurban Kuliev[1] who had just graduated the Turkmen branch of the Moscow conservatory was appointed the main conductor of the orchestra. It opened a new page in the history of the orchestra of the Turkmen national musical instruments. During the years of his guidance a colossal work on perfection of the cast and repertoire of the collective had been done. Thanks to the vigorous creative activity of the orchestra the country learnt new names of musicians and singers who executed solo parts in new works created by K. Kuliev and the others. They were the dutar player Sapar Mjamiev, the honored artist of the USSR and Turkmenistan, gidzhak player Annageldy Dzhulgaev, the vocalists Maya Kulieva, Annagul Annakulieva, Hodzhav Annadurdyev, Margarita Faradzheva, kemancha player A. Kerimov, the tar player A. Bagirov and many others. The tours with their participation had a great success. K. Kuliev guided the orchestra about 30 years. For this time a whole galaxy of musicians and orchestral players who further made a great contribution to the formation of the orchestra of the Turkmen national musical instruments had been brought up.

Besides the aforesaid, the orchestra at last got the opportunity to go on tour outside the territory of the republic. Mainly it was possible thanks to its various repertoires. By 1955 it included more than 20 works of different authors. These were the adaptations of the Turkmen national melodies by G. M. Arakeljan, such as "Lejli gelin", "Njatiljar", the Caucasian sketches "In a village" by M. M. Ippolitov-Ivanov, the "Suite" by K.Kuliev and S.Mamieva's, musical acts from operas "Lejli and Medzhnun" and "Shasenem and Garip", the "Kamarinsky" by M. I. Glinka's, the overture to the 4th act of the opera "Carmen" by J. Bize, adaptation for the tar of the "Concert for domra with orchestra" by A. Budashkin's and many others. Thus the orchestra conducted an active creative and educational activity.

It is necessary to notice that the main merit of Kurban Kuliev as the head and the main conductor of the national orchestra consist in the addition of such musical instruments of a symphonic orchestra as flute, oboe, violoncello, contrabass and timpani into the structure of the Turkmen national orchestra. Certainly, the purpose was justified; due to this measure the repertoire became much wider and more interesting. Nowadays these musical instruments are the part of the orchestra as before.

In 1982 Bajram Hudajkuliev was appointed the head of the collective. He was only about 40 years old, but by that time he had been already a well-known choragus, conductor and composer. B. Hudajkuliev continues the traditions of his predecessor, filling up the repertoire of the orchestra with new, interesting works. Songs by D. Ovezov, V. Muhatov, A. Kuliev, Ch. Nurymov become popular in listeners thanks to the constant tours of the philharmonic cast across the country.

The state had highly appreciated the collective for its fruitful work, and in 1984 the orchestra was appropriated the name of its first head Purli Saryev and rose to the rank of the "Honored" one.

By the end of the eighties one more galaxy of the young composers whose creativity has not avoided the orchestra named after P. Saryev had been brought up. The new creative works by A. Esadov, D. Nuryev, B. Hudajnazarov, N. Muhatov, R. Redzhepov, and D. Hydyrova became the ornament of the repertoire, diversifying it with bright modern colors.

Annageldy Dzhulgaev was appointed the main conductor of the collective in 1986; being the honored artist of Turkmenistan, the composer, conductor and well-known gidzhak player, he guides the

---

[1] Composers and musicologists of Turkmenistan. Ashkhabad. "Turkmenistan", 1982.

orchestra up to now. His appointment was marked for the orchestra by getting of the sharply national sounding. For the purpose of the more picturesque presentation of Turkmen adaptations to the listener, the structure of the orchestra was supplemented through such music instruments as gopuz, gargy tuyduk. Now such melodies as "Jylgajlar", "Baga seyle", "Novajy", "Tuni derja", "Dilberim" or "Ball sajat" sound more intense and affirmative.

Bekmyrat Gutlymyradov who is one of the famous hereditary musicians of Turkmenistan, was appointed the head of the orchestra of the Turkmen national musical instruments named after P. Saryev in 1996. In his turn, this well-known musician and conductor introduced some innovation too: for more saturation of the timbre color and the maintenance of a deep harmonious background he entered accordions into the structure of the orchestra.

In a word, that orchestral structure which is known to us from the sources about the formation of the collective underwent considerable transformation by the end of the nineties; nowadays it is as follows:

Group of wooden wind instruments: flutes, oboes, clarnets, bassoon.

Group of dutars: tonic, second, alto.

Group of plektor dutars: tonic, rubob, baglama (saz), canon, chang, ud.

Bass dutar

Group of percussion instruments: nagara, gosha nagara, deps, shajli dep, small drum, triangle, castanets, timpani.

Group of gidzhaks

Contrabass

It is only a summary, a part of history of the foundation and formation of the orchestra of the Turkmen national musical instruments named after P. Saryev, but the author of the article assumes to finish the monography concerning this theme in the near future.

References

1. Nurymov, J., Goshaeva B. "The Turkmen song and musical art". Ashgabat, 2008.
2. Rezhepov, ☒. "The Honored state orchestra of the national musical instruments named after P. Saryev". "Turkmenistan", Ashgabat, 1992.
3. Uspensky, V., Beljaev V. „The Turkmen music", Moscow, 1928.

APPLIED SCIENCE

Engineering

Autonomous System for Monitoring the Integrity of Navigation Data Provided by Satellite Navigation Systems Based on Optimal Information Processing Algorithms for Navigation Systems of Land Moving Objects

Alexander V. Ivanov, Dmitry Komrakov, Dmitry Boykov

Tambov State Technical University, Tambov, Russia

Abstract. *Optimal algorithms of data processing for the autonomous system for monitoring the integrity of navigation data provided by the satellite navigation system that is a part of the navigation complex of land moving objects were obtained using optimal linear filtration methods. A structural flowchart of the autonomous system for navigation data monitoring was developed. Functional tests of the autonomous system for integrity monitoring were carried out using statistical computer simulation.*

Keywords: *autonomous integrity monitoring; navigation system; land moving object; optimal data processing algorithms; optimization, statistical computer simulation.*

Navigation systems [1] have won wide acceptance for the purposes of tracking the position and identification of trajectory parameters of land mobile objects. Satellite navigation systems (SNS) and inertial navigation systems (INS) (platform or strapdown) are the basis for such systems. Methods of Markov theory of devices and systems integration [2, 3] are widely used for their creation. However, use of a satellite navigation system requires a special algorithm for navigation data integrity monitoring that elicits failures of navigation spacecrafts (NCA) or incorrect data sent by it.

Externally, violation of the integrity of RF signals sent by an NCA manifests itself as signal parameters standing out sharply from a number of measurements. Systems providing for integrity are used to ensure adequate accuracy and reliability. Such systems are usually divided into the following two groups:

– systems using data provided by integrity monitoring facilities that are external with reference to the consumer;

– autonomous integrity monitoring systems onboard of objects.

Responsiveness can be considered as an advantage of autonomous systems. The same principle based on information redundancy underlies their operation. A barometric altimeter (BA) was introduced into the land moving objects navigation system and optimal algorithms for data processing allowing to monitor the integrity of navigation data provided by SNS were synthesized in this paper [4]. It is proposed in this paper to create an autonomous integrity monitoring system that would also use information sent by the inertial navigation system.

Mission: to develop algorithms and a structural flowchart of the system of autonomous navigation data integrity monitoring for navigation systems of land moving objects using additional data from the BA and INS using optimal linear filtration methods.

Task assignment. We will use a barometric altimeter and an inertial navigation system to create an autonomous integrity monitoring system.

We suppose that measurement of altitude with a barometric altimeter is carried out with reference to the level corresponding to the known radius vector R_0 of the geocentric coordinate system, the output signal of the barometric altimeter is as follows [2]:

$$H_{отн}^{ББ}(t_{k+1}) = H_{отн}(t_{k+1}) + \Delta H(t_{k+1}) + E_{6в}(t_{k+1}), \qquad (1)$$

where $H_{отн}^{ББ}(t_{k+1})$ – measured value of the relative altitude; $H_{отн}(t_{k+1})$ – true value of the relative altitude; $\Delta H(t_{k+1})$ and $E_{6в}(t_{k+1})$ – constant component of measurement error of the relative altitude and fluctuation error of the barometric altimeter respectively that can be presented as follows:

$$\Delta H(t_{k+1}) = \Delta H(t_k), \qquad (2)$$

$$E_{6в}(t_{k+1}) = \varphi_u(t_{k+1}, t_k) E_{6в}(t_k) + \gamma_u(t_{k+1}, t_k) n^{ББ}(t_k), \qquad (3)$$

where

$$\varphi_u(t_{k+1}, t_k) = e^{-\alpha_{6в} T};$$

$$\gamma_u(t_{k+1}, t_k) = \sigma^{ББ} \sqrt{1 - \varphi_u^2(t_{k+1}, t_k)},$$

where $\alpha_{6в}$ – ratio that characterizes the error spectrum width; $\sigma^{ББ}$ – mean-root square error of the fluctuation error; $n^{ББ}(t_k)$ – independent samples of Gaussian process with zero mathematical expectation and unit variance.

We presented the signal about the altitude of the object with reference to the center of the Earth at the output of SNS receiving equipment at discrete instants of time in the following way [4]:

$$H^{СРНС}(t_{k+1}) = H_{отн}(t_{k+1}) + R_0 + \sigma_y^{СРНС} n_y^{СРНС}(t_{k+1}), \qquad (4)$$

where $H^{СРНС}(t_{k+1})$ – measured value of the relative altitude; $H_{отн}(t_{k+1})$ – true value of the relative altitude; R_0 – radius vector of the geocentric (spherical) coordinate system; $\sigma_y^{СРНС}$ – mean-root square error in measurements of the object location coordinates; $n_y^{СРНС}(t_{k+1})$ – independent samples of Gaussian process with zero mathematical expectation and unit variance.

We used the principle of data distribution [2] to set the model of object movement and replaced the true value of vertical acceleration by the value measured by the INS, i.e. we used the INS output signal as a component of the control vector that can be presented as follows [5]:

$$a_y^{ИНС}(t_{k+1}) = a_{yg}(t_{k+1}) + \Delta_{ay}(t_{k+1}) + g + \sigma_a\sqrt{\frac{2T}{\alpha_a}}n_{ay}(t_{k+1}), \tag{5}$$

where $a_y^{ИНС}(t_{k+1})$ – measured value of vertical acceleration; $a_{yg}(t_{k+1})$ – true value of vertical acceleration; $T = (t_{k+1} - t_k)$ – sampling interval; g – gravitational acceleration; α_a and σ_a –ratio that characterizes the error spectrum width and fluctuation error dispersion respectively; $n_{ay}(t_{k+1})$ – independent samples of Gaussian process with zero mathematical expectation and unit variance; $\Delta_{ay}(t_{k+1})$ – constant component of acceleration measurement error:

$$\Delta_{ay}(t_{k+1}) = \Delta_{ay}(t_k). \tag{6}$$

Setting the mathematical model of object movement presupposes description of changes in its coordinates and trajectory parameters in the course of time. Changes in coordinates in the course of time can be set using a differential system of equations of the following type:

$$\frac{dH_{отн}(t)}{dt} = V_{yg}(t), \frac{V_{yg}(t)}{dt} = a_{yg}(t). \tag{7}$$

In the mathematical model (7) we switched from the system of differential equations to the difference equation system. Then we used the principle of information distribution [2] to set the model of acceleration changes and replaced the value of vertical acceleration measured by the INS, i.e. we used the INS output signal as a component of the control vector (5). As a result, we got the following for discrete instants of time:

$$\begin{aligned}
H_{отн}(t_{k+1}) &= H_{отн}(t_k) + TV_{yg}(t_k) + 0{,}5T^2 a_y^{ИНС}(t_k) - 0{,}5T^2\Delta_{ay}(t_k) - \\
&\quad -0{,}5T^2 g - 0{,}5T^2\sigma_a\sqrt{\frac{2T}{\alpha_a}}n_{ay}(t_k); \\
V_{yg}(t_{k+1}) &= V_{yg}(t_k) + Ta_y^{ИНС}(t_k) - T\Delta_{ay}(t_k) - Tg - T\sigma_a\sqrt{\frac{2T}{\alpha_a}}n_{ay}(t_k); \\
\Delta H(t_{k+1}) &= \Delta H(t_k); \\
\Delta_{ay}(t_{k+1}) &= \Delta_{ay}(t_k).
\end{aligned} \tag{8}$$

The state vector subject to evaluation includes four components and can be presented as follows:

$$X_в(t_{k+1}) = [H_{отн}(t_{k+1}), V_{yg}(t_{k+1}), \Delta H(t_{k+1}), \Delta_{ay}(t_k)]^T,$$

and in accordance with (8) can be described using the difference vector-matrix stochastic equation:

$$X_в(t_{k+1}) = \Phi_{xxв}(t_{k+1}, t_k)X_в(t_k) + \Phi_v(t_{k+1}, t_k)V_x(t_k) + \\
+ \Gamma_{xв}(t_{k+1}, t_k)N_{xв}(t_k), \tag{9}$$

where $V_x(t_k) = [a_y^{ИНС}(t_k), g]^T$ – known control vector; $\Phi_{xxв}(t_{k+1}, t_k)$ – 4x4 fundamental matrix with nonzero elements $\phi_{xxв_{11}} = \phi_{xxв_{22}} = \phi_{xxв_{33}} = \phi_{xxв_{44}} = 1$, $\phi_{xxв_{12}} = T$, $\phi_{xxв_{14}} = -0{,}5T^2$, $\phi_{xxв_{24}} = -T$; $\Phi_v(t_{k+1}, t_k)$ – 4x2 transfer control matrix with nonzero elements $\phi_{v_{11}} = 0{,}5T^2$,

$ɸ_{v_{12}} = -0.5T^2$, $ɸ_{v_{21}} = T$, $ɸ_{v_{22}} = -T$; $\boldsymbol{\Gamma}_{xв}(t_{k+1}, t_k)$ – 4x2 transfer disturbance vector with nonzero elements $\Gamma_{xв_{11}} = -0.5T^2 \sigma_a \sqrt{\frac{2T}{\alpha_a}}$, $\Gamma_{xв_{21}} = -T\sigma_a \sqrt{\frac{2T}{\alpha_a}}$; $N_{xв}(t_k) = n_{ay}(t_k)$ – forming standard Gaussian random variables.

The observation vector $\boldsymbol{Y}_в(t_{k+1}) = [Y_{в1}(t_{k+1}), Y_{в2}(t_{k+1})]^T$ includes observations at the output of the BA $Y_{в1}(t_{k+1}) = H_{отн}^{ББ}(t_{k+1})$ and SNS reception equipment $Y_{в2}(t_{k+1}) = H_{отн}^{СРНС}(t_{k+1})$ that can be presented as follows in accordance with (1), (4):

$$Y_{в1}(t_{k+1}) = \boldsymbol{H}_{в1}(t_{k+1})\boldsymbol{X}_в(t_{k+1}) + U_в(t_{k+1});$$
$$Y_{в2}(t_{k+1}) = \boldsymbol{H}_{в2}(t_{k+1})\boldsymbol{X}_в(t_{k+1}) + V_2(t_{k+1}) + \Gamma_3(t_{k+1})N_3(t_{k+1}),$$
(10)

where $\boldsymbol{H}_{в1}(t_{k+1})$ и $\boldsymbol{H}_{в2}(t_{k+1})$ – 1x4 observation vectors with nonzero elements $h_{в1_{11}} = h_{в1_{13}} = 1$, $h_{в2_{11}} = 1$; $V_2 = R_0$ – known value; $U_в(t_{k+1}) = E_{бв}(t_{k+1})$ – colored observation noise; $\Gamma_3(t_{k+1}) = \sigma_y^{СРНС}$ – known value; $N_3(t_{k+1}) = n_y^{СРНС}(t_{k+1})$ – observation noise.

We used methods of Markov theory of optimal estimation [2, 3] to get algorithms for complex optimal data processing. The optimum estimate of the state vector is determined as follows:

$$\boldsymbol{X}_в^*(t_{k+1}) = \boldsymbol{\Phi}_{xxв}(t_{k+1}, t_k)\boldsymbol{X}_в^*(t_k) + \boldsymbol{\Phi}_{vв}(t_{k+1}, t_k)V_{xв}(t_k) + \boldsymbol{K}_{в1}(t_{k+1}) \cdot$$
$$\cdot [Y_{в1}(t_{k+1}) - \varphi_u(t_{k+1}, t_k)Y_{в1}(t_k) - \boldsymbol{H}_{в1}(t_{k+1})\boldsymbol{\Phi}_{vв}(t_{k+1}, t_k)V_{xв}(t_k) +$$
$$+\varphi_u(t_{k+1}, t_k)\boldsymbol{H}_{в1}(t_{k+1})\boldsymbol{X}_в^*(t_k) - \boldsymbol{H}_{в1}(t_{k+1})\boldsymbol{\Phi}_{xxв}(t_{k+1}, t_k)\boldsymbol{X}_в^*(t_k)] +$$
$$+\boldsymbol{K}_{в2}(t_{k+1})[Y_{в2}(t_{k+1}) - \boldsymbol{H}_{в2}(t_{k+1})\boldsymbol{\Phi}_{vв}(t_{k+1}, t_k)V_{xв}(t_k) - V_2(t_{k+1}) -$$
$$-\boldsymbol{H}_{в2}(t_{k+1})\boldsymbol{\Phi}_{xxв}(t_{k+1}, t_k)\boldsymbol{X}_в^*(t_k)],$$
(11)

where $\boldsymbol{K}_{в1}(t_{k+1})$ and $\boldsymbol{K}_{в2}(t_{k+1})$ are 4x1 columns of the optimal transfer coefficients matrix $\boldsymbol{K}_в(t_{k+1}) = [\boldsymbol{K}_{в1}(t_{k+1}) \vdots \boldsymbol{K}_{в2}(t_{k+1})]$ that can be determined as follows:

$$\boldsymbol{K}_в(t_{k+1}) = [\boldsymbol{\Phi}_{xxв}(t_{k+1}, t_k)\boldsymbol{P}_в(t_k)\boldsymbol{\Phi}_{yxв}^T(t_{k+1}, t_k) + \boldsymbol{B}_{xyв}] \cdot$$
$$\cdot [\boldsymbol{B}_{yyв} + \boldsymbol{\Phi}_{yxв}(t_{k+1}, t_k)\boldsymbol{P}_в(t_k)\boldsymbol{\Phi}_{yxв}^T(t_{k+1}, t_k)]^{-1};$$
(12)

$$\boldsymbol{P}_в(t_{k+1}) = [\boldsymbol{\Phi}_{xxв}(t_{k+1}, t_k)\boldsymbol{P}_в(t_k)\boldsymbol{\Phi}_{xxв}^T(t_{k+1}, t_k) + \boldsymbol{B}_{xxв}] -$$
$$-\boldsymbol{K}_в(t_{k+1})[\boldsymbol{B}_{xyв} + \boldsymbol{\Phi}_{xxв}(t_{k+1}, t_k)\boldsymbol{P}_в(t_k)\boldsymbol{\Phi}_{yxв}^T(t_{k+1}, t_k)]^T,$$
(13)

where $\boldsymbol{P}_в(t_{k+1})$ – 4x4 matrix of estimation error variance (covariance); $\boldsymbol{\Phi}_{yxв}(t_{k+1}, t_k)$, $\boldsymbol{B}_{xxв}$, $\boldsymbol{B}_{xyв}$, $\boldsymbol{B}_{yyв}$ are block matrixes that can be presented as follows:

$$\boldsymbol{\Phi}_{yxв}(t_{k+1}, t_k) =$$
$$= \begin{bmatrix} \boldsymbol{H}_{в1}(t_{k+1})\boldsymbol{\Phi}_{xxв}(t_{k+1}, t_k) - \boldsymbol{H}_{в1}(t_{k+1})\varphi_u(t_{k+1}, t_k) \\ \boldsymbol{H}_{в2}(t_{k+1})\boldsymbol{\Phi}_{xxв}(t_{k+1}, t_k) \end{bmatrix};$$

$$\boldsymbol{B}_{xxв} = [\boldsymbol{\Gamma}_{xв}(t_{k+1}, t_k)\boldsymbol{\Gamma}_{xв}^T(t_{k+1}, t_k)];$$

$$\boldsymbol{B}_{xyв} = [\boldsymbol{B}_{xy1}|\boldsymbol{B}_{xy2}] =$$
$$= [\boldsymbol{\Gamma}_{xв}(t_{k+1}, t_k)\boldsymbol{\Gamma}_{xв}^T(t_{k+1}, t_k)\boldsymbol{H}_{в1}^T(t_{k+1})|$$

$$\Gamma_{xB}(t_{k+1},t_k)\Gamma_{xB}{}^T(t_{k+1},t_k)H_{B2}^T(t_{k+1})].$$

$$B_{yyB} = \begin{bmatrix} B_{y1y1} | B_{y1y2} \\ B_{y2y1} | B_{y2y2} \end{bmatrix} =$$

$$= \begin{bmatrix} H_{B1}(t_{k+1})\Gamma_{xB}(t_{k+1},t_k)\Gamma_{xB}{}^T(t_{k+1},t_k)H_{B1}^T(t_{k+1}) + \gamma_u^2(t_{k+1},t_k) | \\ H_{B2}(t_{k+1})\Gamma_{xB}(t_{k+1},t_k)\Gamma_{xB}{}^T(t_{k+1},t_k)H_{B1}^T(t_{k+1}) | \end{bmatrix.$$

$$\left. \begin{matrix} H_{B1}(t_{k+1})\Gamma_{xB}(t_{k+1},t_k)\Gamma_{xB}{}^T(t_{k+1},t_k)H_{B2}^T(t_{k+1}) \\ H_{B2}(t_{k+1})\Gamma_{xB}(t_{k+1},t_k)\Gamma_{xB}{}^T(t_{k+1},t_k)H_{B2}^T(t_{k+1}) + \Gamma_3^2(t_{k+1},t_k) \end{matrix} \right].$$

Algorithms (11)-(13) represent complex optimal algorithms for data processing allowing to evaluate the relative altitude and vertical speed component of a land moving object.

We also used the obtained estimate of the state vector to create an autonomous system of SNS navigation data integrity monitoring. To monitor the integrity of SNS navigation data we will use the estimate of the constant component of BA error $\Delta H^*(t_{k+1})$ and the estimate of the constant component of INS acceleration measurement error $\Delta_{ay}^*(t_{k+1})$. Under conditions of SNS normal operation the estimates $\Delta H^*(t_{k+1})$ and $\Delta_{ay}^*(t_{k+1})$ will tend to some constant values determined by the types of BA and INS. In case of SNS failures or artificial transfer of erroneous information from the navigation spacecraft the estimate values $\Delta H^*(t_{k+1})$ and $\Delta_{ay}^*(t_{k+1})$ start to increase. These properties are the properties that we will use to monitor the integrity using the following methods:

– setting the limits ΔH_{max} for the value of BA permanent error estimate $\Delta H^*(t_{k+1})$ and $\Delta_{ay_{max}}$ of the value of estimate of the constant component of acceleration measurement error $\Delta_{ay}^*(t_{k+1})$ of the INS;

– if $|\Delta H^*(t_{k+1})| \geq \Delta H_{max}$ and $|\Delta_{ay}^*(t_{k+1})| \geq \Delta_{ay_{max}}$, data received from the navigation spacecraft cannot be used.

The structural flowchart of the autonomous system for integrity monitoring developed in accordance with the synthesized algorithms (11)-(13) is shown in Figure 1.

Use of these two estimates allows to eliminate the risk of taking any decisions on the integrity of SNS navigation data in case of BA failure, consequently, the quality of decisions increases.

Functional tests of the algorithms that allow to control the integrity of the navigation data were carried out by means of statistical computer simulation. Two instances were considered. In the first instance, as a result of failure the altitude according to SNS data became 100m lower than the altitude according to BA data, while in the second instance, the altitude according to SNS data became 100m higher than the altitude according to BA data.

Simulation of the signal $H_{отн}^{BB}(t_{k+1})$ at the BA output described by equations (1)-(3) was carried out considering the following initial data: $\Delta H(t_k) = 5$ м; $T = 0{,}02$ с; $\sigma^{BB} = 1$ м; $\alpha_{бв} = 10$ с$^{-1}$. It was assumed that the object was moving at an altitude exceeding the radius vector of the geocentric (spherical) coordinate system R_0 by 1000m.

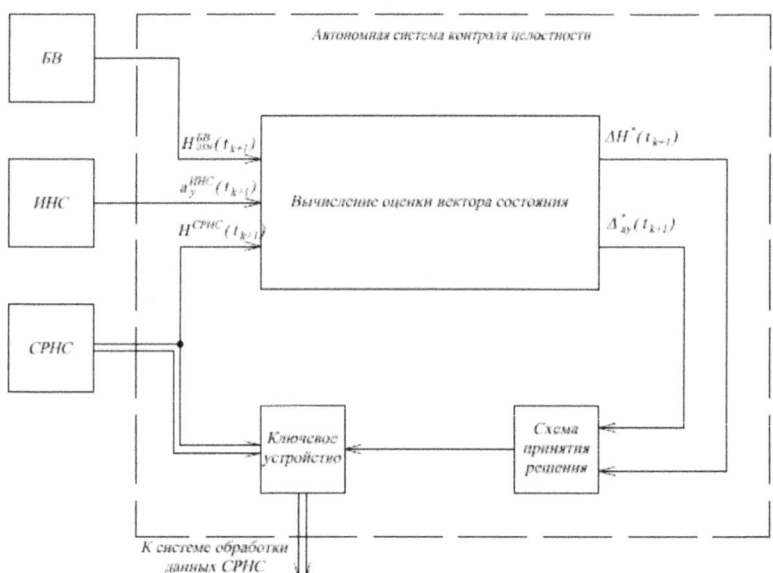

Автономная система контроля целостности – Autonomous system for integrity monitoring
БВ – BA
ИНС – INS
СРНС – SNS
Ключевое устройство – Key device
Схема принятия решения – Decision circuit
К системе обработки СРНС – To the SNS processing system
Вычисление оценки вектора состоянии – Calculation of the state vector estimate

FIG. 1. **Structural flowchart of the autonomous system for integrity monitoring**

Simulation of the signal $H^{CPHC}(t_{k+1})$ at the output of SRS signals reception equipment described by equation (4) was carried out considering the following initial data: $R_0 = 6371110$ м; $H_{отн}(t_{k+1}) = 1000$ м; $\sigma_y^{CPHC} = 3$ м.

Equation (12) was used to calculate the matrix of optimum transfer factors. The initial values of variance of estimate errors of state vector components were equal: $p_{в_{11}}(t_0) = 300$ м$^{-2}$; $p_{в_{22}}(t_0) = 20$ м^{2}с$^{-2}$; $p_{в_{33}}(t_0) = 625$ м$^{-2}$; $p_{в_{44}}(t_0) = 0{,}01$ м^{2}с$^{-4}$.

Figure 2 displays the following: implementation of signal $H_{отн}^{ББ}(t_{k+1})$ at the BA output (Fig. 2, a), implementation of signal $H^{CPHC}(t_{k+1})$ at the output of SNS reception equipment (Fig. 2, b), implementation of the estimate of the constant component of the barometric altimeter error $\Delta H^*(t_{k+1})$ (Fig. 2, c) and implementation of residual error $\Delta(t_{k+1})$ (Fig. 2, d). The diagrams depict the situation when there's a failure in navigation spacecraft operation in the tenth second and there's an error in the navigation spacecraft coordinates that can result in the fact that the relative altitude of the aircraft according to SNS data is 900m but not 1000m (Fig. 3, b).

Figure 3 displays similar diagrams but they depict the situation when as a result of failure the relative altitude of the object according to SNS data is 1100m but not 1000m (Fig. 3, b).

Judging by the abovementioned implementations, it is evident that navigation spacecraft failure can be determined by changes in the estimate $\Delta H^*(t_{k+1})$ of the constant component of barometric altimeter error (Fig. 2, c and Fig. 3, c) and a spike in residual error $\Delta(t_{k+1})$ (Fig. 2, d, Fig. 3, d).

The error of deflection of the relative altitude of a moving object sent by the SNS from the true value, 1000 meters in our case, results in an increase in the estimate of the constant component of barometric altimeter error ΔH^*. This allows to confirm violation of navigation support integrity in case of overriding the threshold limit ΔH_{max} that takes into account maximum possible errors due to propagation of radio waves from the navigation aircraft to the moving object.

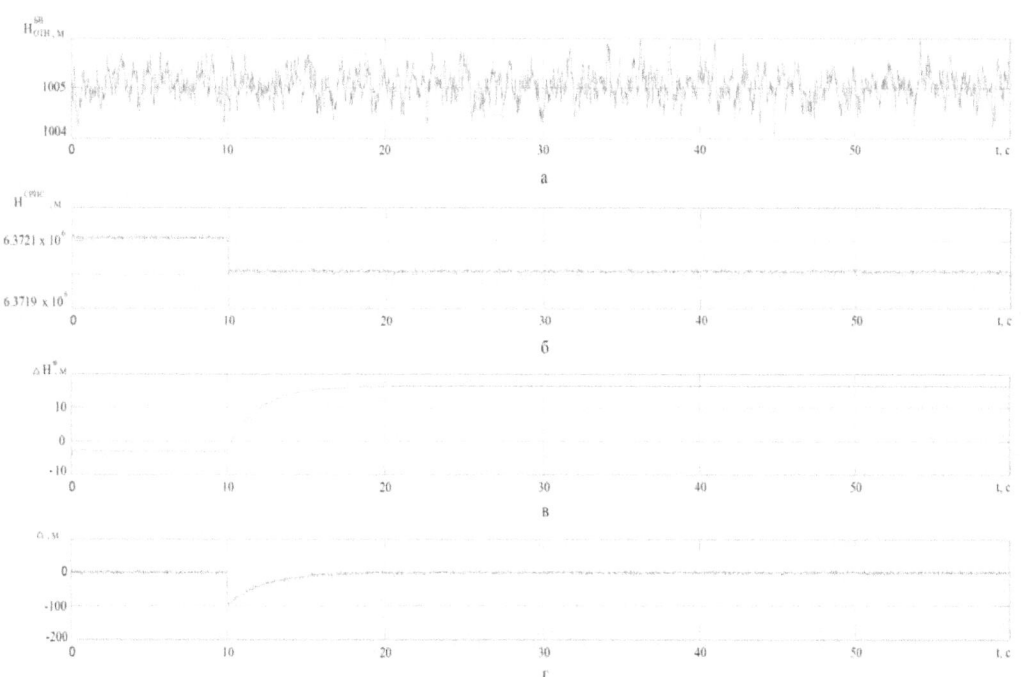

FIG. 2. Results of computer simulation in case of a drop in the relative altitude from 1000m to 900m

When considering these instances, it is evident that errors of deflection of the relative altitude of a moving object sent by the satellite navigation system from the true value result in an increase in the estimate of the constant component of barometric altimeter error ΔH^*. This allows to confirm violation of navigation support integrity in case of overriding the threshold limit ΔH_{max} that takes into account maximum possible errors due to propagation of radio waves from the navigation aircraft to the moving object. Statistical computer simulation also showed that the constant component of acceleration estimate error Δ_{ay}^* of a land object can be used to determine violation of navigation support integrity as an additional parameter.

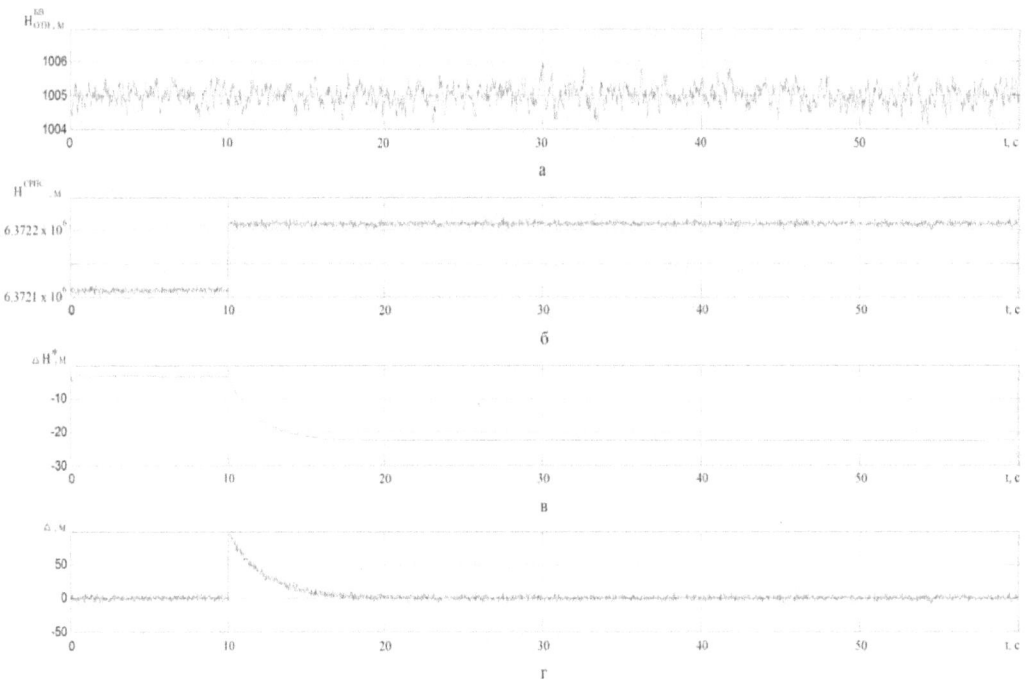

FIG. 3. Results of computer simulation in case of a jump in the relative altitude from 1000m to 1100m

This work is executed within the framework of RFBR grant project "Theoretical foundations of construction of radioelectronic systems equipped with reconfigurable information systems" contract No. NK 14-08-00523/14 of 03/06/14.

References

1. Ivanov, A.V. Navigation of land objects / A.V. Ivanov, N.A. Ivanova. – LAP LAMBERT Academic Publishing, 2013. – 120 p.
2. Yarlykov, M.S. Statistical theory of radio navigation / M.S. Yarlykov. – M.: Radio i Svyaz, 1985. – 344 p.
3. Yarlykov, M.S. Markov theory of estimation of random processes / M.S. Yarlykov, M.A. Mironov. – M.: Radio i Svyaz, 1993. – 464 p.
4. Ivanov, A.V. Complex optimal algorithms for data processing in navigation systems of land moving objects with the control of navigation support integrity/ A.V. Ivanov // Radiotehnika. – 2010. – No.12. – p. 15 – 20.
5. Babich, O.A. Data processing in navigation complexes / O.A. Babich. – M.: Mashinostroyeniye, 1991. – 512 p.

Thermal Insulation Materials in Designing Communications-Electronics Equipment for Hypervelocity Vehicles

Yevgeny Rodikov
Saint-Petersburg State University of Aerospace Instrumentation, Saint-Petersburg, Russia

Abstract. This article provides an analytical review of thermal insulation materials that can be used in instrumentation designed for hypervelocity vehicles.

Key words: thermal insulation materials, aerogel, ATM-10, Solimid, resilient polyimide foams.

Designing of communications-electronics equipment for hypervelocity vehicles issues a standard challenge to draw off heat from device components and another challenge - to protect a flight vehicle against aerodynamic heating. The temperature in the vicinity of the instrumentation module can rise up to 1500°C. And despite the fact that instrumentation modules do not contact directly with such temperatures, the primary thermal insulation sheathing does not provide full protection against heat, as a result it is required to use advanced low-conductivity lightweight and small-size insulation materials to maintain temperature conditions of on-vehicle equipment.

Aerogels

Aerogel is a material that fully complies with the above parameters. It is gel which liquid phase is completely replaced by gaseous phase. The density of such materials is very low, due to it thermal conductivity and weight of aerogels are the lowest.

Aerogels were first discovered by chemist Steven Kistler from College of the Pacific in Stockton (California, USA). He carried out a number of experiments and published their results in "Nature" journal in 1931. Kistler's experiments involved replacing gel fluid by methanol and further gel heating till the temperature of methanol rises up to the critical value of 240°C. Methanol evaporated from the gel but the volume of gel did not decrease, as a consequence the gel structure changed but the volume was the same.

Silica aerogels are best suited for the insulation of communications-electronics equipment for hypervelocity vehicles. Their density yields just the density of metal microlattices, if we speak about solids, which density is up to 0.9 kg/m^3 and it is a shade less than the best density indices of aerogels - 1 kg/m^3. The density of a microlattice in air under normal conditions is 1.9 kg/m^3 due to intralattice

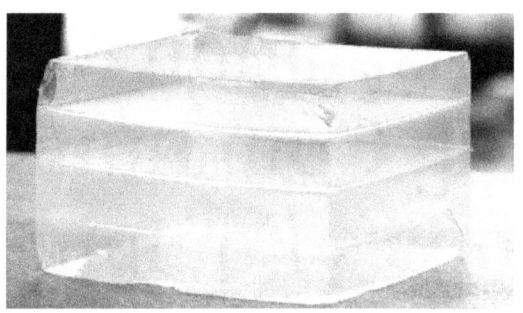

FIG. 1 Appearance of the aerogel discovered by Kistler

air. It is 500 times lower than the density of water and just half as high as the density of air. Due to their extremely low conductivity (~0.017 W/(m•K) in air at atmospheric pressure), that is lower than the conductivity of air (0.024 W/(m•K)), silica aerogels are used as heat retaining or thermal insulation materials. The melting point of silica aerogels is 1200°C, moreover, they are firehard and can be used in aviation industry.

Aerogels are successfully used in aviation and space industry. For example, aerogel is used in Stardust probe. Special types of aerogel for the aviation industry are under development today. For example, aerogel by Aspens Aerogel. This material is best suited for maintaining temperature conditions of communications-electronics equipment for hypervelocity vehicles due to its low conductivity, non-combustibility and small weight.

ATM-10

Russian scientists created thermal insulation material called ATM-3. It looked like a mat consisting of a loose layer of very thin chopped strands covered by glass cloth and stitched by glass fibers. It was used in aircraft manufacturing since it withstands temperatures from -60°C to +400°C. Recently it has been replaced by updated ATM-10 that can withstand high temperatures from -260°C to +900°C. This material represents a mat consisting of basalt very thin chopped strands covered by glass or basalt or silica cloth stitched by fibers.

Materials of which ATM-10 is made make it non-flammable, fire-proof, fire-resistant, environmentally friendly (it contains no phenol binder), vibration resistant, and decay-resistant. That's why

FIG. 2 Aerogel by Aspens Aerogel

TABLE 1. Specification of aerogel by Aspens Aerogel

Thickness	5 mm
Max. utilization temperature	1200°C
Thermal conductivity	0.017 W/(m·K)
Density	180 kg /m³
Hydrophobic properties	Water-proof
Flammability	Non-flammable

ATM-10 is an excellent thermal insulation material for hypervelocity vehicles. ATM-10 is distinguished by large dimensions and it is a significant drawback – it cannot be used for insulation of instrumentation. Currently ATM-10 does not comply with modern requirements to sound insulation, weight, thermal conductivity (>0.06 W/mK), thickness (≥10mm) and fabricability.

FIG. 3 ATM-10

TABLE 2. ATM-10 specification

Thickness	10 mm
Max. utilization temperature	900 °C
Thermal conductivity	0.06 W/(m·K)
Density	43 kg /m³
Hydrophobic properties	Permeable to water
Flammability	Nonflammable

Resilient polyimide foams

Resilient polyimide foams (RPF) are one of the latest developments in the field of thermal insulation (RPFs represent unique materials distinguished by low density and thermal conductivity, high sound absorption, fire resistance, a wide range of operating temperatures, flexibility, and chemical resistance).

TABLE 3. Comparison of polyimide foam plastics

	Solimid	VPP-1
Density, kg/m^3	7–10	7–10
Thermal conductivity, W/(m*K) at 150°C	0.072	0.051
Operating temperatures, °C	from -190 to +200	from -60 to +250
Flammability	firehard	firehard
Resilience	resilient	resilient

A thermal insulation material consisting of resilient foam polyimide sheets one side of which is covered by a polyimide film called "Solimid" by Jenier (USA) is based on RPF. As for fabricability (flexible), density (<0.2 kg/m^3), thermal conductivity (0.05 W/(m·K)) and thickness "Solimid" is second to none domestic materials. Thermal insulation materials based on "Solimid" are used in virtually all Boeing airplanes; Airbus company started to use it. It is also used for thermal insulation of air conditioning systems of RRJ-100 (Russia).

Development of a thermal insulation material based on nonflammable resilient foam polyimide provides for its widespread use in creation of different structures of contemporary and advanced aircraft, in designing surface vessels and submarines, thermal generation stations and nuclear power plants. In Russia (Federal State Unitary Enterprise "All-Russian Research Institute of Aviation Materials") RPFs have been developed since 2009 when semipolymer polyimide powder was selected as a semi-finished product developed by Institute of Macromolecular Compounds Russian Academy of Sciences. An experimental technology of production of nonflammable resilient polyimide foam VPP-1 has been developed recently.

This analytical review of thermal insulation materials for communications-electronics equipment for hypervelocity vehicles shows that currently there are all necessary materials complying with the stringent requirements of the aviation industry (low thermal conductivity, low weight and small dimensions). It allows designers to use them for instrumentation development. Today the most promising and technologically advanced material is VPP-1. It is distinguished by the lowest thermal conductivity and it matches other important criteria set for thermal insulation materials.

References

1. Kablov Y. N. Strategic directions of development of materials and treatment technologies till 2030 // Aircraft materials and technologies. 2012.
2. Goryacheva Y. P., A. Y. Merkuliev Example of maintaining temperature conditions in modern transmitting-receiving devices. // Young Scienticst. 2014.
3. Mikhailin Y. A. Thermal-resistant polymers and polymeric materials. // SpB: Profession. 2006.

Control of Network Standby Power Supply Enabled Once the Islanding System Opens the Sectionalizing Circuit-Breaker

Leonid Surov, Ilya Phillipov, Igor Fomin
Orel State Agrarian University, Orel, Russia

Abstract. *The article provides a description of a method allowing network control that enables standby power supply with the islanding system, a structural diagram and a description of its operation, including representation of output signals.*

Keywords: *power transformer, automatic transfer circuit breaker, operating current decrease sensor, operating current increase sensor, under-voltage relay, recorder*

Rural distribution networks are long and multi-branched. In some cases they allow ring networks (Fig. 1). Sectionalization of the lines forming such a network makes it possible to reserve separate line sections. Reserved sections are connected by means of network automatic transfer circuit breakers, which close in response to loss of voltage on either side. [1].

Loss of voltage in sectionalized line W_1 can occur due to a permanent short circuit in the primary line section, for example, at point K, or due to islanding system operation installed on circuit breaker Q3 (Fig.1) when voltage drops below the allowable value. This article describes a method [2] designed to obtain information about network circuit breaker closing when the busbar circuit breaker opens.

This method allows control of reduction of operating current in the primary power source line by the value determined by the load of the line section adjacent to the network automatic transfer circuit breaker. Moreover, some time after a delay in actuation of network automatic load transfer switch protection, operating current in the standby power supply line is expected to increase by the value by which it decreased in the primary power source line. If these conditions are present, the inference can be drawn that the network reserve is enabled when the islanding system of the ring network switch section is actuated.

For the purposes of implementation of this control method, a structural diagram (Fig. 1) was developed. It consists of the following elements: operating current decrease sensor (OCDS) 1, MEMORY 2, DELAY 3, UNIVIBRATOR 4, operating current increase sensor (OCIS) 5, UNIVIBRATOR 6, I 7, and recorder (RU) 8.

The structural diagram operates as follows. In normal operation of the network, switches Q1, Q2,

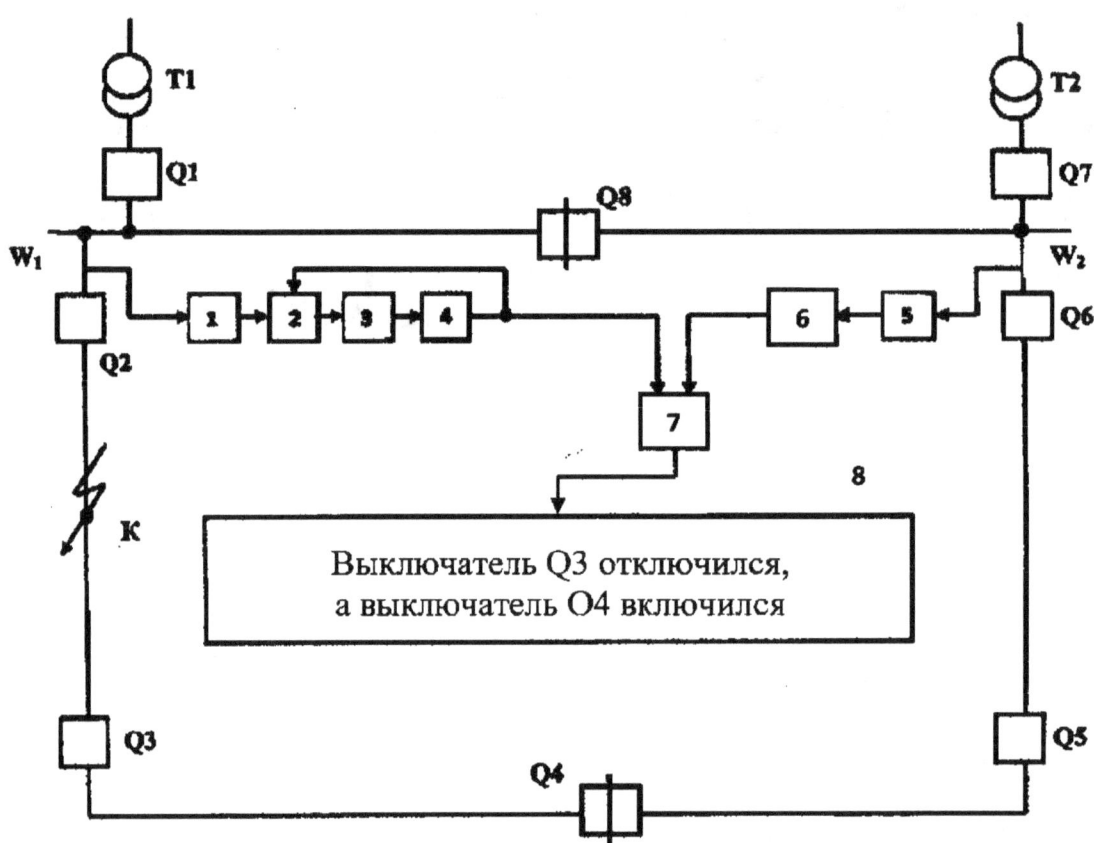

FIG.1. Simplified one-line diagram of a substation consisting of two transformers and structural diagram: T1, T2 — power transformers: Q1, Q7 — isolating switches; Q2, Q6 — primary switches; Q3, Q5 — sectionalizing circuit breakers; Q4, Q8 — network and busbar automatic transfer circuit breakers, respectively

Q3, Q5, Q6 and Q7 are enabled and switches Q4 and Q8 are disabled. There's no signal on the outputs of OCDS 1 and OCDS 5 (Fig. 2, Graph 1), and that's why the structural diagram is in the control mode.

If voltage on the islanding system unit of Q3 switch station drops down to the operating value of the under-voltage relay, it will operate and sectionalizing circuit breaker Q3 will open. It will also open it there is a short circuit in the primary section of line W1, for example, at point K. Opening of circuit breaker Q3 will result in a decrease of operating current in line W1 and sending an output signal to OCDS 1 (Fig.2, graph 1, time point t1) that will reach the input of MEMORY 2, be memorized by it (Fig. 2, graph 2) and then be sent to the input of DELAY 3. The signal will leave this element in a given time equal to the time of delay of the automatic transfer circuit breaker off switch Q4 (Fig.2, graph 3, time point t2) and go to the input of UNIVIBRATOR 4. It will produce one oscillation (Fig.2, graph 4), and the signal will reset the memory from element 2 (Fig.2, graph 2) and go to the first input of element I 7. At this time point (t2), circuit breaker Q4 will close. This will result in an increase of operating current in line W2 by a value equal to the value by which current in line W1 decreased. OCDS 5 will operate; its output signal (Fig.2, graph 5) will go to the second input of I 7. It will operate (Fig.2, graph 7). Its output signal will go to RU 8 and it will receive information that sectionalizing circuit breaker Q3 is open and circuit-breaker Q4 for the network automatic transfer circuit breaker is closed (Fig.2, graph 8).

Implementation of this structural diagram allows obtaining information about enabling network

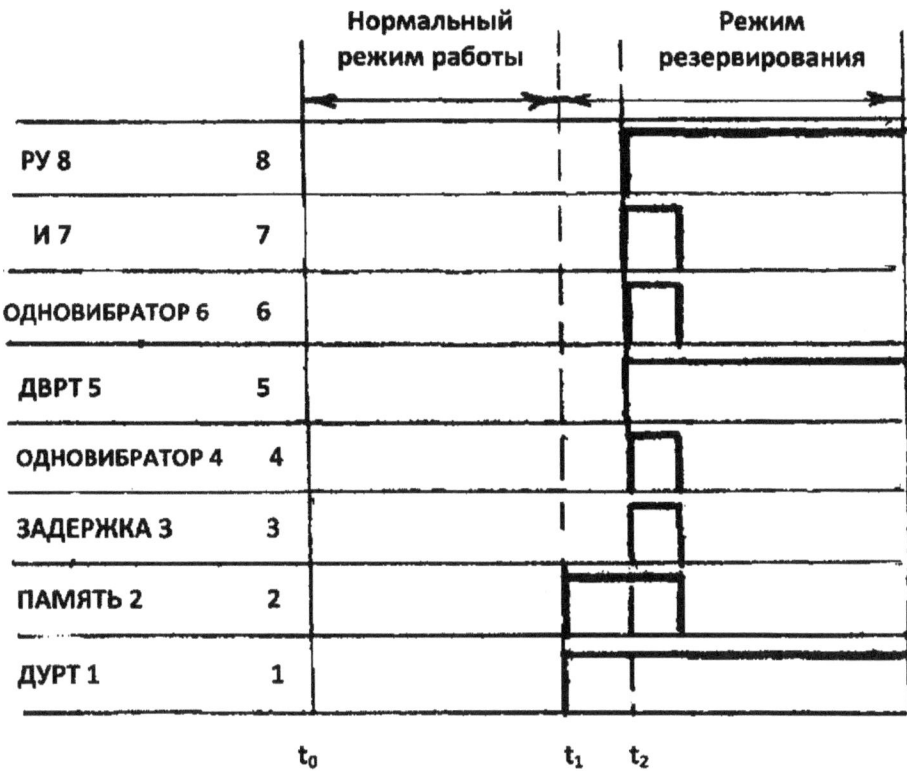

FIG.2. Graph of structural diagram output signals: t_0 — zero time; t_1 — time point when the islanding system operates and circuit-breaker Q3 opens; t_2 — time point when automatic transfer circuit breaker Q4 closes

standby power supply after operation of the islanding system of the sectionalizing point of the ring network line. It allows maintenance personnel to sum up the situation and take all necessary measures to recover power supply and reduce economic losses due to undersupply of energy.

References

1. Automation of electrical power systems: manual for graduate students (Alexeev O. P., Kozik V. A.) — M.: Energoatomizdat, 1994.
2. Patent No. 2542751. L. D. Surov's method to control enabling of network standby power supply once the islanding system of the sectionalizing point of rong network line operates / Surov L. D., 2014.

Medicine

The Choice of Pathogenetic Treatment of Chronic Generalized Periodontitis

Andrey Sushchenko, Olga Oleynik, Elena Vusataya, Oksana Krasnikova, Elena Alferova

Voronezh State Medical University named after N.N. Burdenko, Voronezh, Russia

Abstract. *Pathogenetic features of determined disease forms prognosticate the use of differential approach in treatment depending on the activity of pathologic process in paradontium.*

Keywords: *chronic generalized periodontitis, diagnostics, pathogenetic therapy.*

Introduction

Inflammatory diseases of paradontium (gingivitis and periodontitis notedly) still remain the most commonly encountered forms of pathology [2, 4, 6]. Results analysis of the second national epidemiological dental examination of different Russia regions population which was carried out in 2007-2008 and based on unified criteria recommended by WHO showed that incidence of paradontium tissue involvement in 15-year-old teenagers is 41%, among adult population of 35-44 years it is 18%, in people of 65 and older – signs of the disease are noted practically in all examined patients [5]. Inflammatory diseases of paradontium lead to significant decrease of dental and maxillomandibullar functional capacity with a characteristic prolonged period of rehabilitation. Chronic, recurrent course of this pathology is accompanied by intoxication, sensibilization and decrease of anticontagious organism protection, development of complications, and increased risk of systemic diseases which is an additional argument for special attention to prevention and treatment of inflammatory periodontal diseases [5, 6].

The most severe and widely spread disease is chronic generalized periodontitis (CGP). Except for high incidence, CGP is often associated with inner systems and organs pathology and is characterized by low efficiency of local treatment proving the urgency of this problem [1, 6, 7]. Indisputable fact is that the course of the disease and the effectiveness of its treatment are influenced by preventive measures, early detection and complex treatment. Polyetiology of inflammatory periodontal diseases is well-known; however the emphasis in CGP treatment is still upon elimination of so-called biological factors of the disease development. The investigations of pathogenic microflora are carried out intensively inducing the use of antibiotics in combination with instrumental methods in treatment of paradontium tissue. However, this method is associated, in particular, with pathogenic flora resistance, and antibiotics side effects which influence the organism negatively. Thus, antibiotic therapy cannot be considered a pathogenetic approach for gaining an optimal effect in paradontium treatment. In medical sources one can find a lot of publications concerning research in the field of local and general immunity in patients with paradontium pathology, there are attempts to correct patients' immune response while treatment [2,3,4,9]. Complex therapy is the base for the present-day approach to rehabilitation of patients with inflammatory periodontal diseases. Surgical methods implying the use of pointed regeneration and dental implantation techniques got the most intensive development, but in many cases they don't solve the bacterial problem of paradontium for a long period of time; besides, these techniques don't always contribute to the optimal effect of paradontium bony tissue regeneration affecting only soft tissues. Moreover, the indications for surgical treatment can be determined only visually and by X-ray.

In spite of this, numerous investigations show no decrease in CGP occurrence, initial diagnosis reveals high incidence of severe periodontits (up to 40%) making the failures of diagnostics, treatment and prevention obvious [2, 3, 4, 6]. This fact is due to the absence of integral approach implying the use of conservative pathogenetic therapy in treatment of periodontal diseases.

The investigation objective is: to improve periodontitis treatment effectiveness through development of pathogenetic approach to the choice of therapy methods.

Materials and methods

In the series of clinicoanamnestic, radiological, biophysical, cytological, bacterioscopic, immunological, macro- and microhistochemical investigations we studied the samples obtained in the course of examination and treatment of 365 patients ill with different forms of periodontitis (189 male patients – 52% and 148 female patients – 48%). For the purposes of histochemical and biophysical investigations we studied bioptic samples obtained during curettage, gingivectomy, shifted teeth extraction, and stomatoscopic control of 72 patients (20%). These methods of examination and investigation enabled us to determine the etiological factor in the disease development, as well as links of the pathogenetic chain [1,8,9].

We used the classification of periodontitis which was approved by the All-Union Dentists Society Plenary session in 1983.

Clinical studies included determination of Fedorov-Volodkina genetic index, papillary-alveolar index, Russel parodontal index, as well as methods of visual examination and palpation which determined hundred-per-cent indications for stomatoscopy, bacterioscopy and cytology. We applied both a simple gingevoscopy (stomatoscopy) involving parodontium tissue examination by means of a stomatoscope or a magnifying lens with 10-40 fold magnification, and an extended one involving 4% acetic acid applications (to determine vascular response), 2% Lugol's iodine solution (to reveal foci of inflammation), 1% water solution of nuclear stains (to reveal inflammation and the level of epithelial cell mitotic activity). The depth of parodontal pockets was determined by the graduated probe.

The first stage of clinico-laboratory investigations was mouth fluid pH determination which is one of the oral cavity homeostasis main indicators with the help of «ORATEX - 3001».

In the course of biophysical study in order to reveal tissue viability and to predict treatment results, we determined the value of gum polarization efficiency.

Bacterioscopic investigation involved mucous membrane and periodontal pocket scrapings; staining and microscopic study were performed by A.A. Kunin technique (1976) as well as by Romanovskiy-Gimze and Gram ones. Preparations for cytologic investigation were necessary to study the number and maturity level of epithelial cells, leucocytes, lymphocytes, macrophages under immersion with 7x90 magnification.

To reveal immune system dysfunction in case of periodontitis we studied both organism cellular and humoral immunity: T- and B-lymphocytes, blood serum immunoglobulin M, A, G (L.A. Malinovskaya, 1987) as well as oral cavity local immunity: immunoglobulin A in oral secretion, saliva lysozyme (S.V. Yerina, 1989). Control group consisted of 50 people with an intact paradontium.

Clinico-laboratory diagnostics included complete blood cell count and feces analysis in order to reveal gastro-intestinal tract dysbacteriosis reflected in the composition of oral cavity microflora.

Samples for the histochemical investigation, after mucous membrane particles fixation in Karnua mixture and paraffin embedding, were arranged to study ploidy value of mucous membrane epithelial cells nuclei, ribosomal RNA topography, and to determine total protein content.

Ethical principles stated in Helsinki declaration of 1964 and modified by 41st World Assembly in Hong-Kong in 1989 were complied in the course of clinical investigations.

Results

The highest incidence of periodontitis is observed in people at the age of 35-54 (85.8%). The number of patients with a mild form of the disease is 17.2%.

On the basis of clinico-laboratory investigations analysis we were able to reveal 2 groups of patients. 167 people (45.5%) with slowly progressing chronic course of generalized periodontitis, moderately marked hyperemia of gum mucous membrane, recurrent purulent discharge from parodontal pockets, gradually appearing mobility of some teeth with periods of pathologic process stabilization. Congestive inflammatory type of response was observed during a simple gingevoscopy; an extended one - showed epithelium blenching (in case of 4% acetic acid use). Shiller-Pisarev test revealed light brown staining of the gum, nuclear stains gave light blue gum color. Immunological investigations gave evidence of T- and B-lymphocytes as well as T-suppressors functional ability disorder in these patients, blood serum had high inhibitory ability. Reliable increase of serum immunoglobulins M, G, A (348.5 ± 31.4; 350.6 ± 60; 321.9 ± 6.57 IU/ml) was revealed in comparison with a control group (162.7 ± 13.9; 184.1 ± 18.7; 190.4 ± 27.2 IU/ml). Functional activity of T-lymphocytes was reduced from the very beginning of the disease (60.7 ± 4.4%) by contrast to healthy T-lymphocytes (70.4 ±2.5%; $P < 0.05$). Functional activity of T-suppressors was reduced (56.1 ± 3.7%) compared to the control (74.8 ± 4.8%). Natural inhibitory factor of blood serum increased with disease progress (up to 3.8 ± 0.8%) comparing with the control (1.1 ± 0.8%). In the passive course of periodontitis, epithelium morphological changes were characterized by insignificant modifications of several epithelial cells, limited acanthosis, epithelial layer thinning. RNA histochemical investigations in this group of patients showed extinction decrease up to 2.8.

The second group of patients (198 people –54.5%) had an active continuously progressing course of the disease. Intensive persistent hyperemia was determined by clinical tests; in some cases hyperplastic processes were revealed. Mobility of teeth appeared and progressed quickly due to increasing destruction of interalveolar septum. Bright red colour of the gum mucous membrane, marked swelling, minute vessels dilation gave evidence of an acute inflammation course during a simple gingivoscopy. Epithelium blanching, red-brown and dark-blue gum staining were revealed during an extended stomatoscopy with the use of relevant reagents. During repeated immunological investigations blood serum did not show inhibitory activity, inhibition index remained less than 1.1, T-lymphocytes functional activity (66.5 ± 41) was slightly different from the control (70.4 ± 2.5) and it didn't decrease as the disease progressed. Lymphocytes had functional hyperactivity. Characteristic features of the samples obtained from patients ill with frequent periodontitis exacerbations were extensive acanthosis, dystrophic changes of epithelial cells in the spinous layer, parakeratosis. In patients with the active periodontits course a typical structure of an epithelial layer was absent, it was extremely thin, and epithelium proliferated into connective tissue. Histochemical study of DNA and RNA, total protein in an epithelial layer in case of the active periodontits course showed increase of RNA extinction up to 2.85-2.9. In 70% of cases there were tetra- and hexaploid epithelial cells nuclei, total protein content in stratified squamous gum epithelium increased. We revealed an important biophysical factor characterizing the level of gum viability - polarization efficiency (A.A. Kunin, 1998). If marked chronic inflammatory reaction of a hyperplastic gum had a low response to administered earlier conservative treatment and the polarization efficiency was less than 1.5, there were indications for gingivectomy (141 patients,

38.6%) as the level of studied tissue viability was very low and did not prognosticate positive results of conservative therapy. In other patients with polarization efficiency of 1.5-1.9, the level of oxidation-reduction processes in gum tissue and absence of cell membranes damage defined by this biophysical factor value gave prognosis for good result in case of anti-inflammatory low intensity laser therapy preceding operative treatment. Other patients (168 - 46%) had polarization efficiency of 2.0 and higher defining a hyperergic type of metabolic processes which gave indications for flap operations and easy mucous membrane graft with the initial anti-inflammatory inhibiting laser therapy. Cytologic investigation was of particular importance in determination of pathologic process character in parodontium. In 335 patients (97.2%) we revealed significant number of both blood corpuscles (leucocytes, lymphocytes) and epithelial cells with different maturity level. New cells with nuclear-cytoplasmic ratio 1:1 – 1:3 were revealed in patients with an active inflammatory process. Passive processes in the gum were characterized by mature epithelial cells forms with nuclear-cytoplasmic ratio 1:2 – 1:3.

To reduce the possibility of dental plaque formation we studied oral fluid pH in order to choose a tooth paste. 135 patients (36.9%) had a neutral reaction (6.9-7.1), 162 patients (44.3%) had an acidic reaction (<6.9), in other 68 people (16.1%) – the reaction was alkaline (>7.1). To gain positive results of hygienic measures we recommended patients to use toothpastes with an adequate pH level: in case of neutral saliva – «Colgate» or «Colgate total – maximum caries protection», for patients with acidic saliva – «Colgate – soda bicarbonate», patients with alkaline saliva were recommended to use «Colgate F + Ca».

Results of patients examination just after the treatment showed positive changes in values studied in 83% of patients with the passive periodontitis form and in 69% of patients with the active one.

Discussions

It is at the age of 35-40 years that periodontal diseases begin to progress and, eventually, lead to pathologic mobility and development of tooth alignment defects. Much fewer number of patients with a mild periodontitis form indicates late diagnosis and gives evidence not only of rapid process progression but also of first inflammation signs early manifestation.

According to our approach of pathogenetic paradontium diseases treatment it is inexpedient to use symptomatic means. Histochemical and biochemical investigations carried out by members of our department revealed presence of metabolic disorders in paradontium tissue. We received evidence of significant disorders in T- and B-immunity systems. According to experimental clinical data, natural inhibitory factor (NIF) decreases the activity of different specificity macromolecular antibodies and affects immunogenesis depressively. In conditions of physiology inhibitory features appear as shot-term, but in some pathologic states NIF continues to function over a long period of time contributing to the development of acquired immunodeficiency state.

As our investigations showed, one of the most important pathogenesis links is gum metabolic disorders accompanied by exceeding or reduced DNA or RNA concentration in gum epithelium depending on the process activity, alongside with significant changes in glycogen content. In terms of cytological investigation method it should be noted that if the presence of blood corpuscles can be explained and is not of a diagnostic significance, the determination of epithelial cells maturity level showed the activity of the process conforming to macro and micro histochemical investigations data, determining indications and the dose choice of laser periodontitis therapy. Presence of new cells with nuclear-cytoplasmic ratio 1:1 – 1:3 indicated their mitotic hyperactivity and the activity of the inflammatory process prognosticating the use of anti-inflammatory inhibitory laser therapy doses. Chronic gingivitis was characterized by mature epithelial cells forms with nuclear-cytoplasmic ratio 1:2 – 1:3 indicating regenerative epithelial activity.

Investigation of oral fluid pH in order to choose toothpaste gives a chance to decrease the possibility of dental plaque formation which affects unfavorably parodontium tissues being also necessary for administration of pathogenetically based parodontium treatment.

Thus, clinical, morphological, histochemical and immunological investigations of patients ill with periodontitis made it possible to determine 2 forms of the disease: active and passive.

In this context, patients with the passive periodontitis form were recommended measures to increase reparative regeneration of parodontium tissue in terms of DNA and RNA level decrease in cells - pyrimidine derivatives accelerating cell reproduction and growth processes in parodontium tissue. They include pentoxyl in the dose of 0.2gr x 3 times a day during the period of 2-3 weeks and methyluracil 0.5gr – 2-3 times a day with the therapy course of 30 days. For RNA and DNA synthesis inhibition in patients with the active periodontitis form, we used 30% oily solution of vitamin E (30-40gr) in the dose of 1gr a day orally during the whole course. These patients were also recommended local application of vitamin E solution on wick drains into parodontal pockets every day during the period of 20 days. Vitamin E is actively absorbed by parodontium tissue and leads to long-term absence of process exacerbation.

The results of long-term periodontal diseases investigations let us draw a conclusion that all surgical measures carried out in treatment of this pathology must be based on preliminary normalization of gum mucous membrane viability with the use of high- and low-peak power laser radiation. Our opinion is that to gain positive results of surgical therapy methods it is necessary to have an appropriate level of parodontium tissue viability – it mustn't be low (p.e. – 1.5) or too high (p.e.– 2 and higher, hyperergic type). The obtained results enable future investigations of pathologic processes in parodontium.

Conclusion

1. Scientific approach to diagnosis and treatment of periodontal diseases revealed significant priority of complex of methods obligatory implementation giving the possibility to determine pathogenetic aspects of the pathologic process. According to our approach of pathogenetic parodontium treatment the use of symptomatic measures is not expedient. Histochemical and biochemical investigations carried out by the members of our department showed presence of metabolic disorders in parodontium tissue accompanied by hyper or hypo concentration of DNA and RNA in gum epithelium alongside with significant changes of glycogen content.

2. Notable disorders of T- and B- immunity systems, decrease of different specificity macromolecular antibodies activity under the influence of NIF affect immunogenesis depressively. Thus, present-day idea of periodontal diseases gives us the chance to determine 3 basic links in their pathogenesis – microcirculatory, metabolic and immunological, complex disorders of which are an essential part of different pathologic processes development (scheme).

3. Pathogenetic features of determined disease forms prognosticate the use of differential approach to treatment depending on pathologic process activity in parodontium. This approach was tested on great number of patients. The choice of offered methods in parodontium diseases treatment is based on positive statistic and the experience of introducing scientific investigations into practice. Pathogenetic approach to periodontitis treatment scheme:

1) normalization of gum disturbed metabolic disorders;

2) normalization of parodontium microcirculation;

3) normalization of the immune status.

4. The results of long-term periodontal diseases investigations let us draw a conclusion that all surgical measures carried out in treatment of this pathology must be based on preliminary normalization of gum mucous membrane viability with the use of high- and low-peak power laser radiation. Thus, laser therapy possessing multiple-factor pathogenetic action is an important link in periodontal diseases prevention and treatment.

5. Treatment of periodontal diseases should be based on the principle of individual attention to each patient taking into account general and dental statuses. It must include: in the first instance – multiple factor therapy directed to elimination of primary factors; second – pathogenetic therapy implying all means and methods of influence on different links of pathogenesis with regard to pathologic process activity.

References

1. Burger, F. Working out of indications and methods and evaluation of high- and low-peak power laser radiation results in periodontitis therapy: synopsis of a thesis, MD F. Burger; medical academy – Voronezh, 1999. – p. 54.
2. Grudyanov A.I. Parodontium Diseases. – Moscow: Medical Information Agency Press. 2009. – p. 26-37.
3. Incidence of dental diseases in Russia. The condition of parodontium tissue and oral cavity mucous membrane/ under the editorship of prof. O.O. Yanushevich. – Moscow: MSUMD, 2008. – p. 10-12.
4. Oleynik, O. I. Development of methods and assessment of effectiveness of individual prevention of inflammatory periodontal diseases: abstract of dissertation of doctor of medical sciences. - Voronezh, 2014. – 46 p.
5. Oleynik, O. I., Vusataya E. V., Popova S. V. Integrated approach to the treatment of early forms of periodontal inflammatory diseases // The Young scientist. – 2015. - № 5 (85). – p. 75-78
6. Periodontics: the national manual / ed. Prof. L. A. Dmitrieva. – M.: GEOTAR-Media, 2013. – 712 p.
7. Periodontitis/ under the editorship of prof. L.A. Dmitrieva. – Moscow: MEDpress-inform, 2007. – p. 54-88.
8. Sushchenko, A. V. Principles and methods of diagnostic research rational correction of clinical and laboratory characteristics of the mucous membrane of the oral cavity in diseases of the digestive tract: abstract of dissertation of doctor of medical sciences. – Voronezh, 2004. – 36 p.
9. Vusataya, E. V. The medical social characteristics of dental morbidity of urban adult population and ways of its improvement in modern conditions: dissertation of candidate of medical sciences.– Ryazan, 2007. – p. 32-34

Current Aspects of Parkinson's Disease Treatment

Vyacheslav Kutashov, Natalia Kameneva, Maxim Yurin, Nikolay Zhuchkov
Voronezh State Medical University named after N.N. Burdenko, Voronezh, Russia

Problem statement

The last decade of the 20th century was proclaimed "The Decade of Brain." Brain diseases attracted considerable interest by researchers in medical science because of the high incidence of the problem. Neurodegenerative diseases attracted special attention, including Parkinson's disease, which was described in 1817 by the English physician James Parkinson.

Parkinson's *disease is the most common neurodegenerative disease; it not only leads to severe neurological disorders, but also significantly reduces patients' social adaptation and quality of life. The occurrence of Parkinson's disease among people older than 70 years is 107 to 187 cases per 100,000. The occurrence of typical morphological characteristics of Parkinson's disease (intracellular Lewy's bodies in the stem structures) significantly exceeds the incidence of symptomatic forms of the disease and, according to previous research, was found to be 5% to 7% of elderly people.*

The increased incidence of Parkinson's disease affects the health and socioeconomic indicators that are specific to diseases of the nervous system. Parkinson's disease leads to physical disability, and increases requirements for special care for patients, leading to higher costs, especially in the tardive dyskinesia stages of the disease.

In *recent years, exploration of the clinical symptoms of the disease and how to cure it have continued. There has been active development of neurochemical and molecular genetic studies that focus on the etiology, pathogenesis and pathophysiology of Parkinson's disease.*

Keywords: *Parkinson's disease, treatment.*

Current aspects of the treatment of Parkinson's disease

The complexity of treatment for Parkinson's disease contributes to the urgency of finding new treatment approaches. Currently, the standard treatment for Parkinson's disease is pathogenetic therapy utilizing levodopa, which is usually used in combination with a peripheral decarboxylase inhibitor(levodopa + benserazide [Madopar], levodopa + carbidopa [Duellin], sindopa, and others.). Use of those medications leads significant changes in the disease, improved quality of life and increased the survival rates.

The present article, rather than considering a specific treatment for Parkinson's disease, focuses on an overview of the most effective groups of drugs that have a reliable evidence base.

Current aspects of the use of the new generation of antiparkinson drugs

Undoubtedly, one of the major achievements in the treatment of Parkinson's disease in recent years has been the widespread use of dopamine receptor agonists. Although this class of antiparkinson drugs has been known since the beginning of the 1970s, it gained practical importance with the advent of a new generation of drugs like pramipexole and ropinirole. It has been shown that in the early stages of the disease dopamine receptor agonists are capable of delaying both the use of levodopa for quite a long period and the time of occurrence of motor complications.

The results of one study showed that the use of pramipexole by patients with Parkinson's disease for a year led to a significant reduction of the severity of motor defects: an average of 20.9% in the early stage, and an average of 19.3% in the tardive stage. Stable adequate therapeutic effects of pramipexole were retained at the end of the year of treatment in 53.5% of patients with Parkinson's disease, including 75% of patients in the early stage and 35.1% of patients in the tardive stage. In the case of the tardive stage of Parkinson's disease, use of pramipexole led to significant prolongation of the effect of single dose of levodopa, by an average of 28%. The combination of pramipexole and levodopa was able to reduce the dose of levodopa by an average of 11.1%. Also, pramipexole led to a decrease of 18% in the severity of depressive symptoms as assessed by Beck's scale, mainly due to the reduction of anxiety, depression and somatoform symptoms. To avoid the side effects of pramipexole, special attention should be paid to elderly patients with severe cognitive disorders and concomitant cardiovascular diseases. For long-term pramipexole treatment, body weight control is recommended [1].

Current aspects of treatment of disease cognitive disorders in Parkinson's disease

Cognitive disorders in Parkinson's disease must be mentioned. Thus, it has been shown that in the tardive stages of the disease, it is reasonable to use the neuroprotector memantine and the cholinomimetic galantamine, which have been shown to be highly effective. With akatinol memantine or galantamine monotherapy, as well as combined therapy, cognitive disorders were significantly reduced in a group of patients, mainly due to improvement in neural and regulatory functions (attention, logical memory, verbal activity) and improvement of psychomotor functions. Significant improvement in association tests (36.6%) after akatinol memantine treatment was found. However, improvements were insignificant after galantamine treatment. Treatment with both drugs caused a significant decrease in emotional disorders and improved quality of life for patients in the tardive stage of Parkinson's disease. However, akatinol memantine therapy led to greater regression in degree of anxiety than galantamine therapy [#]. In a comparison of the therapeutic effect of the two drugs, akatinol memantine and galantamine in patients in the tardive stage of Parkinson's disease, memantine has been shown to have an advantage for improving motor function [#].

It is necessary to point out that in the case of insufficient effectiveness of galantamine, akatinol memantine monotherapy is recommended for long term treatment with maintenance doses. In the case of insufficient effectiveness of akatinol memantine or galantamine as monotherapies, a combination of akatinol memantine and galantamine is suggested. To reduce the severity of side effects during the initial period of treatme, the dosage of the drugs should be titrated gradually, and in the presence of gastrointestinal disorders, it is recommended that domperidone be used [2].

Current aspectsthe use of neuroprotectors for treatment of Parkinson's disease

The inclusion of neurometabolic drugs, such as karnitin and emoxypine is justified in complex drug therapy for patients with Parkinson's disease. Use of these drugs provides improved cognitive function, reduces the severity of extrapyramidal and other neurological disorders, can slow the rate of disease progression, and provides the possibility of reducing the dosage of antiparkinson drugs, particularly medications that contain L-dopa. It is necessary to pay attention to synthetic antioxidants-mexidants. It has been shown that in Parkinson's disease, mexidants neutralize the growth of lipid hydroperoxide levels and increase antioxidant protection. The most clearly proven mexidant effect is reducing the frequency of detectable side effects of levodopa therapy [3].

The prescription of the neuropeptide carnosine for Parkinson's disease has shown a significant increase in the positive dynamics of symptoms when compared with a group of basic therapy drugs. Carnosine decreases the oxidative damage of lipoproteins of blood on the back of higher levels of endogenous antioxidant protection and provides preservation of the antioxidant enzymatic activity of superoxide dismutase. It helps to improve the overall motor activity, decreases the severity of hypokinesia, rigidity and tremor, as well as improves daily activity tests. [4]

Current aspects of PK-Merz use

On the tardive stages of Parkinson's disease, during long term Levodopa therapy there is an increase in the risk of developing the most serious complications, such as akinetic crisis, which is manifested by high mortality. The most effective for combined therapy of Parkinson's disease akinetic crisis is found to be the drug PK-Merz (active substance: amantadine, a blocker of glutamate NMDA-receptors). Thus, the result of therapy, including preparation PK-Merz was cupping crisis, reducing the severity of a disease in a group of patients studied by 27%, decrease the severity of dysphagia (45%) and akinesia (26%), improvement of daily activity (87%). For relief akinetic crisis it is necessary to use a parenteral PK-Merz (amantadine sulfate), instantly dispersible Madopar, because in the presence of dysphagia oral use of drugs is difficult [5].

Conclusion

Taking into account the information above, it is necessary to cure Parkinson's disease with a certain regimen. In addition to pathogenic agents, such as dopamine receptor agonists (Pramipexole, Ropinirole), and levodopa, it includes symptomatic therapy, which is aimed to improve cognitive functions, and acetylcholine drugs (Akatinol, Memantine), that have a positive influence in the occurrence of complications - akinetic crisis (PK-Merz). Only complex use of those drugs can stabilize the patient with this pathology and achieve positive changes in the treatment of Parkinson's disease, carrying a huge social and economic role in the 21st century.

References

1. O.S. Levin, A.N. Boiko, O.S. Nesterova, O.V. Otcheskaya, E.YU. Zhuravleva. Effect of dopamine agonist pramipexole (mirapex) on tremor, affective disorders and quality of life in patients with Parkinson's disease. S.S. Korsakov Journal of Neurology and Psychiatry. 110(2): 39-44.
2. Kozhevnikova, Z. 2009. Differential approach in the treatment of cognitive disorders in the later stages of Parkinson's disease [dissertation]. [Moscow]: Institute of professional development of Federal medicobiological agency of Russia.
3. Kuznetsov, N. 2010. Parkinson's disease and vascular parkinsonism: differential diagnosis and treatment [dissertation]. [Moscow]: Russian State Medical University of Federal Agency of Healthcare and Social Development.
4. Bagyeva, B.H. 2010. Clinical, genetic and biochemical analysis of Parkinson's disease: mechanisms of predisposition, experimental models, and approaches to therapy. [dissertation] [Moscow]: Scientific center of neurology of the Russian Academy of Medical Science.
5. Chigir, I.P. 2006. Approaches to the treatment of Parkinson's disease decompensation. [dissertation for candidate of medical science] [Moscow]: Russian Medical Academy of Postgraduate Education
6. Yahno, N. 2010.Dementia: a guide for physicians. Moscow:"Medical press inform"

Current Peculiarities of Diagnosis and Treatment of Benign Paroxysmal Positional Vertigo

Vyacheslav Kutashov, Valentina Matviets

Voronezh State Medical University named after N.N. Burdenko, Voronezh, Russia

Abstract. *In clinical practice, doctors of many specialties find patients who complain of dizziness [1]. The incidence of dizziness in ambulatory neurology, general practice and otolaryngology varies from 10% to 30%. Every year, dizziness is observed in 4.9% of the population [8]. It has been shown that 9% of elderly patients who consulted a doctor about other problems revealed benign paroxysmal positional vertigo (BPPV) upon examination. In the United States, 7.5 million patients per year complain of vertigo in general practice and emergency care, and this is one of the most common reasons for seeking medical consultation [6].*

Keywords: *Benign paroxysmal positional vertigo (BPPV), systemic vertigo, Dix–Hallpike test, Pagnini-McClure test, Brandt-Daroff exercises, Epley maneuver, Semont maneuver.*

Statement of the problem

A uniform standard for examination of vertigo patients has not been developed to date. Patients often receive ineffective treatment due to wrong diagnoses. In most cases, vertigo is mistakenly interpreted as a manifestation of vertebrobasilar insufficiency, vascular encephalopathy, or disease of the cervical spine. It often takes several months to make a correct diagnosis. Even though the indispensable examinations are carried out—computer (CT) and magnetic resonance imaging (MRI) of the head, cervical spine x-ray, and ultrasound examination of the carotid and vertebral arteries—ineffective treatments are still prescribed. All of this slows the recovery of patients [3]. Although the symptoms of vertigo are widespread, diagnosis of the diseases causing dizziness shows significant difficulty due to misunderstanding of the term "vertigo." Vertigo occurs in lesions of the central vestibular system (strokes, multiple sclerosis, migraines) or at the peripheral level (benign paroxysmal vertigo, Meniere's disease, and psychogenic dizziness) [4]. There are many different methods for assessing the state of the vestibular system, from calorimeter testing and videonystagmography to MRI and electrocochleography. However, in most cases, the cause of vertigo can only be diagnosed based on a thorough analysis of patient complaints and medical history, as well as the results of a simple neurological and otoneurological examination. Careful collection of medical history allows determination of the cause of dizziness in 75% of cases [5].

The separation of vertigo into systemic and non-systemic is clinically significant. Systemic vertigo is the illusion of body movement or movement of environmental objects. All other sensations in this type of balance disorder, such as instability, unsteadiness when walking, feelings of faintness, blackouts, and lightheadedness are not associated with pathology of the vestibular apparatus and are called non-systemic vertigo. Vertigo caused by disorders of the peripheral vestibular apparatus has a sudden beginning and end, the seizure duration is limited (usually no longer than 24 hours) and combined with severe autonomic disorders (nausea, vomiting); during the interictal period patients feel well. There is also rapid activation of central compensatory mechanisms and residual vestibular dysfunction lasts no longer than 1 month. Dizziness due to impairment of the central vestibular system is not clearly defined. It has a chronic course, duration is not clear (days, weeks, months), it is combined with symptoms of central nervous system damage [4].

Pathogenesis of BPPV

Benign paroxysmal positional vertigo is the most common cause of peripheral vestibular vertigo [6]. The average age of occurrence BPPV is from the fifth to seventh decade of life [3]. This condition was first described by Adler in 1897. In 1921, Barany provided as more detailed description. It is a disease of the labyrinth, which manifests in episodes of positional vertigo and prevails in the elderly, commonly women. In most cases (70%-80%) the cause of the disease cannot be determined (idiopathic BPPV). In some cases, the cause may be cranial trauma, surgery of the middle ear, labyrinthitis, stapedectomy, respiratory infections, suppurative otitis media, or alcohol and barbiturate intoxication. Idiopathic cases of the disease are in most cases related to the degenerative process and the formation of deposits in the cupula of the anterior semicircular canal. As a result, there is an increase in a spinning sensation by a direct gravitational effect on the cupula or by inducing endolymph flow during head motion in the direction of gravity. Development of BPPV can be explained by the formation of otoliths (otolithiasis) in the cupula (cupulolithiasis) or the semicircular canals (canalolithiasis). Frequently (60%-90%) the disease process is located in the posterior semicircular canal, and more rarely in horizontal and anterior canals [5]. The otoliths (earstones or otoconia) are deposits of calcium carbonate in the form of composite calcite crystals and contain otolith membranes, which serve as a load for the receptors located in the spherical and elliptical bags of the vestibule. Under the influence of various factors (trauma, viral infection, and others), otoliths may move out of the otolith membrane and, with certain movements of the patient, may move in the semicircular canals, resulting in irritation of ampullary receptors. Typically, ampullary external receptors (horizontal) or rear (vertical) of the semicircular canals are damaged. BPPV is accompanied by short episodes of systemic vertigo, which appear in certain positions of the head or body, most often when bending the head forward and down, turning over or lying on the back, tilting the head back, and making the transition from horizontal to vertical. Episodes may occur with some or all of the triggering movements and may last up to 60 seconds. Usually, patients know in what position of the head they may occur. Sometimes vertigo presents when turning on one side or the other, but the intensity of the vertigo is more pronounced on the affected side; it disappears quickly if the patient does not change position. If the position changes several times, the vertigo disappears and does not reoccur, but it may appear again after a long break. Neurological symptoms such as hearing loss or tinnitus are absent [4].

Diagnosis and treatment of posterior semicircular canal BPPV

Diagnosis of BPPV is based on medical history and typical clinical presentation of episodes. Diagnosis can be confirmed by positional tests, including the widely used Dix-Hallpike test, which allows identification of the most common type of BPPV, which is due to lesions in the posterior semicircular canal. The test is conducted as follows: After seating the patient upright, the head is turned 45° in the direction of the involved ear (right or left). The patient is then moved from the sitting to the supine position. The patient's head is held by doctor's hands and hangs over the edge of the examination table in a relaxed state. The sample is considered to be positive in case of presence of vertigo and the appearance of mixed horizontal and rotatory nystagmus with a slight delay. The latency (time from bending until nystagmus) in the pathology of the anterior and posterior semicircular canals is no more than 3-4 seconds, while the duration should be 30-40 seconds [6]. The Dix-Hallpike test must be carried out on both sides. If there is no nystagmus and vertigo during the test, it is considered to be negative. In some cases the test may be positive on both sides; this is especially typical for post-traumatic BPPV. It is necessary to point out that nystagmus, which presents in the test, is definitely peripheral, i.e., it suppressed by eye fixation. Therefore, it cannot be detected without special equipment (Frenzel goggles or infrared recording of eye movement) [7].

Treatment methods for BPPV have changed substantially over the last 20 years due to progress in understanding the pathogenesis of this disease. Previously, patients were advised to avoid triggering behaviors and medication was symptomatic. Later, new techniques and maneuvers were found that allowed otolith fragments to move back into the utricle. In 1988, Semont described a technique based on the theory of cupulolithiasis. He suggested that performing a series of rapid changes of head position would detach the particles attached to the cupula. After seating the patient on the examination table, the head is turned away from the involved ear. Then the patient is quickly placed on the side toward the affected ear, and the head position is maintained, i.e., in this case the patient is in face-up position. After 5 minutes, the patient returns to the initial position on the other side, while the head is still rotated in the "healthy" direction (face down). The patient is left in this position for 10-15 minutes, and then slowly returns to the original position.

According to the Epley maneuver, proposed in 1992, the patient is placed in Dix-Hallpike position, as mentioned above, and remains in this position for 1-2 minutes. Then the patient quickly turns on one side, with the head rotated 180 °, to a position diagonally opposite to the Dix-Hallpike position. If vertigo and nystagmus occur (continued movement of otoconia in an ampullofugal direction towards the utriculus), the maneuver has succeeded. The patient is kept in this position for 1-2 minutes. Then the patient returns to a sitting position, and if the manipulation has succeeded, does not feel vertigo or nystagmus. In the original description, the author used skull vibration to increase efficiency, but following studies showed that the Epley maneuver is very effective regardless of the use of vibration. The effectiveness of the proposed methods is undeniable and has been confirmed by numerous studies [7].

Diagnosis and treatment of horizontal semicircular canal BPPV

The diagnosis of horizontal semicircular canal BPPV can be carried out by using the Pagnini-McClure maneuver. The head is turned about 90° to each side while supine. In the case of canalolithiasis, geotropic nystagmus (downward motion) can be observed. In the case of cupulolithiasis, apogeotropic nystagmus (upward motion) can be observed. Compared with canalolithiasis, the cupulolithiatic type of BPPV tends to have a longer time frame (up to 2 minutes). The lesion side can be determined by the severity of vertigo and nystagmus. For the treatment of horizontal semicircular canal BPPV, a barbecue roll maneuver is widely used. The patient is rotated in the direction opposite to the involved ear, making a 360° turn. Cupulolithiasis treatment is more complicated. Before treatment, the cupula otoliths adhering to the cupula must be released, that they can move freely in the lumen of the canal. There is a variety of possible techniques, and the Gufoni maneuver is one.

During the various maneuvers, canal switching may occur as a complication. It is not an indicator of a wrong technique, and may be related to the anatomical structure of the semicircular canals of certain patients.

If the repositioning maneuvers fail or if patients cannot tolerate the repositioning maneuvers, the Brandt-Daroff exercise may be prescribed. The patient is instructed to rapidly lie on one side, sit up, lie on the opposite side, and then sit up again, and the head is turned up to 45°. These exercises are repeated in 3 series 5 times a day, until resolution of the symptoms. Then the patient stops the exercise and visits the doctor [2].

According to a study by Bestuzheva et al. (Department of Nervous Diseases and Neurosurgery of the I.M. Sechenov First Moscow State Medical University of the Ministry of Health), higher efficiency of the BPPV treatment utilized, compared other studies, could have been related to the fact that after a therapeutic maneuver, all patients received Betaserc (48 mg/day) for 2 months. Normalization of the vestibular function using betahistine is related to: improvement of blood supply (activation of H1-histamine receptors in the internal ear); reduction of asymmetric operation of the peripheral vestibular sensory apparatus; and effects on the posterior regions of the nucleus of the hypothalamus (increased synthesis of hyper stamina) and the vestibular nuclei of the brain stem (suppression of H3-histamine receptors) [3].

Conclusion

It should be noted that, according to most foreign researchers, BPPV is a separate nosological form, based on escape of the otolith fragments from the utriculus otolith membrane and their movement to the endolymphatic space of the semicircular canals, often adjustable. Clinically it declares itself by brief vertigo (for 20-30 seconds) associated with certain positions of the head and body, sometimes accompanied by nausea, while there are no signs of auditory analyzer. In the most cases it is idiopathic or occurs as a result of head injury. The presence of BPPV with other diseases of the inner ear or central nervous system is most likely a random finding, i.e., in these cases BPPV is a concomitant disease. BPPV is well-diagnosed and can often be cured using simple techniques without any surgical or medical intervention [7].

References

1. Kosivtseva, OV, Zamergrad, MV. 2012. Vertigo in neurological practice: general issues of diagnosis and treatment. Jnl of Neurology, Neuropsychiatry, and Psychosomatics. 1: 48-51. (In Russian)
2. Guseva, AL. Benign positional vertigo: peculiarities of diagnosis and treatment. Department of Otolaryngology of Medical Faculty of the Pirogov Russian National Research Medical University (RNRMU.) NI Pirogov Video Lecture published on Aug. 6, 2015 http://www.youtube.com/ (In Russian)
3. Bestuzhev, NV, Parfenov, VA, Antonenko, LM. 2014. Diagnosis and treatment of benign paroxysmal positional vertigo in ambulatory practice. Jnl of Neurology, Neuropsychiatry, and Psychosomatics. 4: 26-29. (In Russian)
4. Mullayanova, RF, Yakupov, EZ. 2014. Benign paroxysmal positional vertigo: the difficulty of diagnosis. News of Curr Clin Med. 7(2): 123-125. (In Russian)
5. Parfenov, VA, Abdulina, OV, Zamergrad, MV. 2010. Differential diagnosis and treatment of vestibular vertigo. Neurology, Neuropsychiatry, and Psychosomatics. 2:49-54. (In Russian)
6. Sadokha, KA, Fursova, LA, Minnikova, VA. Vertigo: frequent causes, diagnosis. 2015. Medicine. 1(41): 52-56. (In Russian)
7. Baybakov, EV. 2012. Benign paroxysmal positional vertigo: diagnosis and treatment. Otolaryngology (Russian medical journal). 27: 1370-1373.
8. Neuhauser, HK. 2005. Epidemiology of vestibular vertigo. Neurology. 65: 898-904.

Conversion in Laparoscopic Cholecystectomy

Margarita Ryzhikova, Anna Soloviyova
Belarusian State Medical University, Minsk, Belarus

Abstract. *This article provides information about primary reasons for cholecystectomy, its forms, advantages and disadvantages of laparoscopic (closed) and laparotomic (open) access, as well as the stages of laparoscopic cholecystectomy and reasons for its conversion in laparotomic.*

Keywords: *laparoscopic and laparotomic cholecystectomy, conversion, gall bladder.*

Non-invasive and minimally invasive diagnostic and treatment methods are preferred in medicine in the 21st century - the age of high technologies. And surgery is not an exception. Different types of diseases and, consequently, the volume of surgical treatment require the most attenuated for the patient access to be chosen ensuring the shortest term of temporary disability and reducing the number of complications. For example, appendectomy is currently carried out using either the open or closed method depending on the preferences of the surgeon, the patient and available surgery equipment, while laparoscopy has replaced conventional laparotomy for gall bladder removal almost completely.

Despite the fact that gallstones were described in the V century by Greek physician Alexander Trallianos, first cholecystectomy (from Latin chole - bile, cyst - bladder, ectomy - removal) – a surgical technique for gall bladder removal – was carried out by Langenbach in Berlin in 1882. At that time this operation was subjected to enormous criticism due to a large number of complications. In the following years anatomical knowledge and surgical experience accumulated, the number of intra- and postoperative complications reduced and also the number of deaths reduced significantly. It was primarily due to improved diagnosis of gall bladder diseases, while the operation itself did not undergo significant changes until the introduction of laparoscopic cholecystectomy. First cholecystectomy using laparoscopic technique was carried out by Muhe in Germany in 1985. Mouret used a laparoscope to remove the gall bladder in France in 1987. Laparoscopy was not a new operation at that time: it was described in 1901 but till that time it was only used for diagnostics. Development of materials, optics and technologies allowed it to acquire a number of advantages over conventional operation methods over time and begin to replace them.

Doubtless advantages of the laparoscopic access for surgical interventions in the gall bladder are shorter hospital stays as well as faster recovery and getting round to labor activities, better cosmetic results, less severe pain syndrome during the postoperative period and less likely development of an infection and evisceration. But surely, this technique also has disadvantages: a) the image on screen display is not three-dimensional; b) the surgeon cannot carry out examination by touch; c) it is practically impossible to extract gallstones from biliary tracts and, consequently, to a larger or smaller extent, depending on the experience of the surgeon, it is required to proceed from laparoscopic surgery to laparotomic; d) in some cases it is necessary to carry out conversion of closed cholecystectomy in open cholecystectomy; d) it requires specific instruments; e) it requires the surgeon to have certain skills.

When cholecystectomy is required? The main indications for cholecystectomy are as follows: cholelithiasis, gall bladder injuries, gall bladder tumors, cholecystitis (acute and chronic), and intestinal obstruction caused by a gallstone.

Depending on the access cholecystectomy can be open (laparotomic) and closed (laparoscopic).

A large number of laparotomic accesses to expose the liver, gall bladder and bile ducts are described. There are vertical and oblique incisions. The most common incisions are Kocher's and Fyodorov's oblique incisions as they provide the most direct way and the best access to the gall bladder, bile ducts and the visceral surface of liver.

The length of Kocher's incision is 15-20 cm, it begins from the median line and ends 3-4 cm below and parallel to the right costal arch.

The length of Fyodorov's incision begins from the xiphosternum, continues 3-4 cm down the median line and then parallel to the right costal arch; its length is 15-20 cm.

Upper median, pararectal and transrectal approaches are referred to vertical incisions of the anterior abdominal wall. The median incision that is made between the xiphosternum and the omphalus is the most frequently used incision from this group. In case this access is not enough, it can be expanded by making an additional right transverse incision.

Oblique incisions also include angular and undulating incisions by Kehr, Bevan, Rio Branco, Cherni, Braitsev, Mayo Robson, and Kalinovsky that provide a free access to the bile ducts and liver. The most frequently used incision is Rio Branco's incision that is made along the median line from the xiphosternum down and turned right and up to the end of X rib two fingers' breadth up to the omphalus.

Laparoscopic access is provided using special instruments by means of 3-4 punctures in the abdominal wall 5-10 mm in diameter. Special tubes (trocars) are introduced into the punctures. The patient shall be in the Trendelenburg position with the feet higher than the head by 15-20 degrees. The first trocar 10-11 mm in diameter is introduced through a small incision in the upper umbilical fold blindly. The trocar is introduced by rotational movements directing it end towards the pelvic cavity. Upon trocar introduction the obturator is removed, the source of carbon dioxide is attached and the gas is injected through the laparoscope cannula with a camera and a lamp attached to it. First, the surgeon shall examine all visible parts of the abdominal cavity and find injuries that could arise with the introduction of the first trocar. Then an auxiliary cannula 5 mm in diameter is introduced along the anterior auxiliary line at the omphalus level at the right. Nontraumatic forceps holding the gall bladder bottom while turning up the liver will be introduced through this channel. Its lower surface and the gall bladder are also opened at the same time. The third trocar 5 mm in diameter is introduced along the midclavicular line 4-5 cm below the costal arch edge. Other forceps are introduced through this channel and put onto the gall bladder stulk opening Calot's triangle. The fourth trocar 10-11 mm in diameter is introduced into the epigastrium at a distance of 4-5 cm from the xiphosternum and slightly to the right of the median line. Special instruments, such as dissecting probes, spatulas, forceps, aspirators, and irrigators will be introduced through it. Places for trocar introduction can be changed depending on the position of the lower edge of the liver and gall bladder localization. Currently, there is a tendency to reduce the number of punctures at the anterior abdominal wall. [1], [2]

Currently, laparotomic accesses are rarely used during cholecystectomy, that's why this article contains a description of laparoscopic cholecystectomy stages: 1) the position of the patient is on the back, anesthesia, intubation of trachea; 2) laparoscopic access, insufflation (carbon dioxide is commonly used); 3) cholecystectomy (gall bladder and gall ducts mobilization; searching for Calot's triangle elements; clipping of the cystic artery and cystic duct; gall bladder removal); 4) abdominoscopy, instruments extraction, gas removal; 5) closure of skin punctures.

However, it is not always possible to end cholecystectomy through the laparoscopic access. Surgeons often have to change tactics during surgery and carry out conversion of laparoscopic cholecystectomy in laparotomic cholecystectomy. Conversion is a refusal to continue endoscopic surgery and its completion in a conventional way. Due to it the aim of the examination is to analyze the reason for the transition from laparoscopic cholecystectomy to laparotomic.

The main indications for conversion are as follows: 1) anatomic features of each individual patient; 2) change of diagnosis in the course of the operation; 3) morphological changes (Mirizzi syndrome, adhesion, dense infiltrate); 4) complications during surgery (bleeding, internal injuries); 5) patient's massive obesity; 6) equipment failures that often result in complications during surgery. [2] According to BelMAPGE (Belarusian Medical Academy of Post-Graduate Education) available dense infiltrate in the area of the gall bladder is the reason for conversion during laparoscopic cholecystectomy in 64% of cases. [1]

In the course of the study an analysis of medical records of 620 patients of the department of surgery of health care institution "Minsk clinical hospital No.3" who underwent cholecystectomy during 2013-2014 was carried out. 620 (100%) cholecystectomies were laporoscopic, 20 cholecystectomies ended in conversion in laparotomic (representing 3.2% of all cholecystectomies). The main reasons for conversion of laparoscopic cholecystectomy in laparotomic were cystic artery (arteria cystica) bleeding (14 cases) and Mirizzi syndrome (6 cases).

The cystic artery (arteria cystica) is often a branch of the right hepatic artery (arteria hepatica dextra) that in its turn starts from its own hepatic artery (arteria hepatica propria) and the latter one - from the celiac trunk (truncus coeliacus).

Mirizzi syndrome is a complication of gall stone disease in which an inflammatory and destructive process in the gall bladder neck and common hepatic duct develops resulting in the development of bile ducts stenosis and fistula formation.

Thus, currently, cholecystectomy is almost 100% laparoscopic. It is converted in laparotomic in about 3% cases and the main reasons for that are bleeding and Mirizzi syndrome.

References

1. I.N. Grishin, A.V. Vorobey, S.V. Alexandrov. Complications and hazards in laparoscopic surgery as reasons for transition to the "open" operation method./ Medicine. - 2006. – No.4.
2. A.F. Popov, A.S. Balalykin. Reasons for conversion in laparoscopic surgery. / Endoscopic surgery.- 1997.- No.1.- P.87.

Osteoplastic amputation in the middle third of the thigh

Ryzhikova Margarita, Soloviyova Anna
Belarusian State Medical University, Minsk, Belarus

This article contains information about basic indications for amputations, techniques of their carrying out, as well as information about using the osteoplastic method to form stump in the middle third of the thigh.

Keywords: *osteoplastic amputation, middle third of the thigh.*

Over the past few decades due to the emergence and introduction of high technologies into all spheres life, the way of living of world population has drastically changed. It results in a steady increase in metabolic diseases (mainly, diabetes) that lead to changes in blood vessels and, consequently, histotrophic and organotrophic nutrition. Limb vascular diseases are the leading cause for amputations in people aged 50+ years (90% of all amputations in this age group). Amputation is one of the most mutilating surgeries that results in work decrement up to incapacitation and inability to adapt to everyday life. For example, amputation of a distal phalange of the fifth digit of a non-functioning arm is characterized by 5% persistent disability; amputation of a forearm results in the third group of disability; available prosthetic stumps of both legs – in the second group of disability; and short stumps of both thighs - in the first group of disability. Therefore, this operation requires the surgeon to have a clear knowledge of indications and selection of the most favorable for a particular patient amputation techniques.

Amputation (from Latin *amputare – to cut, remove*, from Latin *ambi - everywhere* and from Latin *putare – to clip, cut*) is removal of a devitalized distal extremity segment along a bone or bones.

Amputation is one of the oldest surgeries. Hippocrates carried out amputations on dead tissues and Celsus - on healthy tissues. In the 16th century, Ambroise Paré introduced ligation of intersecting vessels during amputation. In 1720, Jean-Louis Petit reconstructed the method of covering the bone stump by a skin cuff, and in the 19th century Nikolay Pirogov suggested osteoplastic operations during which the bonesaw-line is covered by bone autologous graft (free or cut out with a flap of soft tissues).

The most difficult issue in amputation is determination of indications for its carrying out. According to MRED statistics (medical rehabilitation expert board) the most common causes for amputation are traumas - 48%; vascular diseases - 42%; tumors and congenital deformities - 10%.

All indications for amputation are commonly divided into invariable and relative.

The first group of indications includes gangrene of soft tissues caused by burns, freezing injuries, electrical injuries, diabetic angiopathy, anaerobic infections, endarteritis, embolism, and injuries of extremities that resulted either in its traumatic avulsion or a triad - two-thirds of soft tissues of the extremity are damaged, major neurovascular tracts are damaged or crushed and there's a complete transverse bone injury. Their invariability is attributable to processes inconvertibility and inefficiency of non-surgical therapy aimed at preserving the extremity continuity.

In case of relative indications, the issue related to extremity amputation is considered on an individual basis taking into account the condition of a particular patient in each separate case. The operation is usually performed when the trauma or disease of the extremity is life-threatening to the patient.

Relative indications also include malignant tumors; unremediable congenital, paralytic and post-traumatic extremity deformities; persistent, resistant to non-surgical treatment and steadily progressive extensive trophic ulcers; extremity injuries when two-thirds of all soft tissues are crushed and a significant segment of the bone is damaged provided that the continuity of neurovascular tracts is preserved.

Stages and procedure of amputation: [1]

The position of the patient is on the back, the extremity to be operated is on the side table. Anesthesia: general.

1. Apply an arresting bleeding tourniquet. It allows to transect all soft tissues with no blood. Many authors do not recommend to apply a tourniquet due to increased ischemia in the ischemic extremity.

2. Dissection of soft tissues and subcutaneous tissue. By type of soft tissue incision amputations are as follows:

1) guillotine amputations are as follows: a) linear amputations, when all tissues are transected at the same level (in case of life-threatening infections); b) one-stage amputations — after skin incision soft tissues and the bone are transected along the border of its displacement; c) two-stage amputations when tissues are transected at the same level up to fascia, then after retraction of transected tissues muscles are proximally transected and the bone is sawed at the level of displaced muscles; d) three-stage (conical-circular) amputations (Pirogov's method) — fascias, superficial and deep muscles and the bone are transected at different levels;

2) flap method — the most widespread method of amputation. Single- and double-flap amputations exist. The wound is covered by one or two flaps respectively. Flaps are formed from skin and subcutaneous fat. If the flap includes fascia, this type of amputation is called fascioplastic. In most cases the length of a long flap shall be equal to 2/3 of the diameter of the extremity at the amputation level, and its width – to the full diameter of the extremity at the amputation level. The length of a short flap shall be equal to 1/3 of the diameter of the extremity at the amputation level, i.e. half of the length of a long flap. Thanks to it the cutaneous scar of the stump is displaced from the end of the stump to its non-functioning part and it facilitates subsequent prosthetics. The best way of flaps cutting out is when the scar is located on the back surface of leg and thigh stumps.

3. Incision of muscles.

4. Bone and periosteum treatment. By the way of periosteum treatment the following bone treatment methods are distinguished: subperiosteal, aperiosteal and transperiostal. The subperiostal method presupposes periosteum dissection distal of the bone cut level and its proximal movement to cover the bonesaw-line after bone sawing off by the periosteum. In practice, this method can only be applied to children due to good elasticity of their periosteum. The aperiosteal method presupposes periosteum dissection 0.5 cm proximal of the intended level of bone cut and its distal stripping. It is almost impossible to move the periosteum without damaging it in adults and damaged areas become the places where bone spurs – osteophytes - grow henceforth making the stump unsuitable for prosthetics. Currently the transperiostal method is usually used to treat the bone during amputation. Moreover, the bone is sawed in close proximity to the dissected periosteum, 1-2 mm distal of its edge. [3]

5. Treatment of vessels. Prior to tourniquet removal ligatures are applied on all large vessels in the stump, moreover, two ligatures are applied on arteries and the lower one shall be pierced: one of its ends is passed through the needle which is used to sew both walls of the artery. This additional fixation prevents ligature from slipping. To prevent ligature fistulas all vessels are tied up by threads from absorbable materials and muscle vessels are pierced. Arterial and venous vessels are tied up separately. They are carefully separated from the perivascular tissue using surgical forceps. Fin blood vessels are coagulated. [4]

6. Treatment of nerves (alcohol block). Nerves are transected, at least, 5-6 cm proximal of the level of amputation. Nerve trunks that are not truncated as required can result in creation of nerves matted together with the stump scar tissue, that's why the nerve shall be carefully separated from the surrounding tissues and transected in one movement by a safety razor. First, a perineural injection of 3-5 ml of 2% novocaine solution with 1 ml of 96% alcohol solution (procaine block) is administered.

7. Layered closure and stump formation.

Stump formation methods: a. cutaneous-fascial - the sawline is covered by a flap of skin, subcutaneous tissue and fascia; b. tenontoplastic - the sawline is covered by muscle tendons; c. osteoplastic - the sawline is covered by the other part of the bone; [2] d. myoplastic - antagonists are sewn together over the sawline.

In our study we pursued the aim to analyze the reasons for choosing and the course of osteoplastic amputation in the middle third of the thigh based on the data provided by the septic surgery department of healthcare facility "Minsk municipal clinical hospital No.2" collected during 2011-2014. We analyzed data of 320 medical records of the septic surgery department patients.

89 amputations were carried out in 2011, 54 - in 2012, 81 - in 2013, 96 - in 2014 of which 47 (53%), 36 (67%), 59 (73%) and 63 (66%) in 2011, 2012, 2013, 2014, respectively were amputations in the middle third of the thigh. In 2011 and 2012 osteoplastic amputations in the middle third of the thigh were not carried out. 13 amputations (22% of amputations in the middle third of the thigh and 16% of all amputations) were carried out in 2013 and 21 amputations (33% and 22%, respectively) – in 2014.

According to the septic surgery department of "Minsk municipal clinical hospital No.2" osteoplastic amputation in the middle third of the thigh is carried out in two stages:

1. Separation of a bone transplant from the fibular bone.

2. Amputation in the middle third of the thigh and transplantation of the bone fragment separated earlier into the marrowy canal of the femoral bonesaw-line.

Thus, over half of amputations are carried out in the middle third of the thigh. The septic surgery department of "Minsk municipal clinical hospital No.2" started to carry out osteoplastic amputations in 2013. The tendency towards an increase in the number of cases (an 11% increase in amputations in the middle third of the thigh, a 7% increase in all amputations) was noted in 2014. It is preferred to carry out osteoplastic amputations from the point of view of reduction of the term required to stump healing and provision of better prosthetic possibilities in future.

References

1. Kryzhova Y.V. and others. Amputation of extremities: Method. recom./ Y.V. Kryzhova, S.I.Korsak, A.A.Bayeshko, O.V.Lopukhov. - Mn.: MSMU, 2001. - 11 p.
2. Voronchikhin SI, Vostroknutov VV. Bone plastic hip amputation in endarteritis obliterans and arteriosclerosis. - Ortop Travmatol Protez, 1997. - c. 41-62
3. Baumgartner Rene, Botta Pierre. Amputation and prosthetics of lower extremities./ R. Baumgartner, P. Botta. - Monograph. – Translation from German. - M.: Medicine, 2002. - 504 p.
4. Burakovsky V.I., Bokeriya L.A. and others. Cardiovascular surgery. - M.: Medicine, 1996. - 767 p.

Post-stroke Depression

Alexander Shulga, Vyacheslav Kutashov, Marina Shulga
Voronezh State Medical University named after N. N. Burdenko, Voronezh, Russia

Abstract. *Depression is a common occurrence after stroke and is associated with excess disability, cognitive impairment, and mortality. The authors undertook a review of the literature to review several aspects of this illness, including the prevalence of this disorder and its etiology.*

Keywords: *depression, stroke, reasons of post-stroke depression.*

Depression is related to one of the most common post-stroke disorders. Depression is a mental disorder characterized by pathological low mood, negative and pessimistic self-feeling, negative feeling about one's position in real life and one's future.

Development of depression by stroke patients impairs cognitive functions and the quality of life, increases length of hospitalization period and suicide risk [1].

A basic criterion for diagnosing depression, according to DSMIV, is presence of 5 or more of the following symptoms during the last 2 week-period or more:

— low mood most of the daytime;
— loss of interest or pleasure in almost all activities (anhedonia);
— weight loss;
— insomnia or hypersomnia (sometimes insomnia at night and hypersomnia during the day);
— psychomotor agitation or retardation;
— fatigue;
— feeling of worthlessness or guilt;
— reduced ability to concentrate thoughts;
— recurrent thoughts of death, suicide ideas (or suicide attempts).

If at least the first 1–2 and to 4 following depressive symptoms are present, is diagnosed a minor depressive disorder. If at least the first 2 and 4 (and more) following (accessory) depressive symptoms are present, is diagnosed a major depressive disorder.

After the cerebral stroke depressive disorders are present by 34,8 % patients, their average age is 63,5 and they are mostly women. Depressive disorders occur in different age groups at different rates. The highest depression rate have the most young individuals (under the age of 49) and the most old individuals (over the age of 80). At the age under 59 depressive disorders are most common among women and at the age above 70 — among men. At the age of 60–69 frequency of occurrence of depression is among women and men equal [2].

A post-stroke depression can appear not only during the acute but also during the early post-stroke rehabilitation period [3]. By the patients with the post-stroke depression is mortality rate during the first ten years after the stroke higher in comparison with the patients without the post-stroke depression [4]. The post-stroke depression increases cognitive impairment including spatial orientation difficulties, speech disturbance and pain syndrome intensity [5].

The average length of the post-stroke depression ranges from 6 to 9 months. Additionally, sometimes manifestation of the post-stroke depression symptoms can be observed immediately after the stroke, sometimes delayed, a part of the delayed post-stroke depressions during the acute stroke period — 12 %, in the early and late rehabilitation period respectively 29 and 9 %.

When analyzing frequency of occurrence of depression of different depth, it is estimated, that among the post-stroke depressions prevail minor depressive disorders (82 %), subsyndromal depressive disorders and major depressive disorders tend to be more rarely [6].

A large part of the post-stroke patients is affected by a so called latent (masked or larvate) depression, in which neither the patient has any complaints of lowering mood, no her/his relatives see any significant emotional deviations of the patient. But a careful neurologist can notice that the patient has got symptoms of the latent depression. As these symptoms are considered [7,8]:

— contradictory, numerous and changing complaints;

— mismatch between the patient's complaints and clinical picture of disease and results of paraclinical studies;

— sleep disorder;

— loss of appetite and loss of weight during the last weeks and months;

— feeling of expressed fatigue, which is not connected with physical or mental tension;

— patient's appearance: negligence in dress, poverty of facial expression, glassy look;

— monotone and weak-modulated speech, which is not connected with affection of language zones in the brain.

Among the post-stroke depressive disorders worried depressions are the most common (43 % cases). They begin to appear with despondency and predominance of anxiety about disability, loss of capacity for self-care, about possible death, concerns about future and also about future of relatives and about outcome of current situation. The symptoms get more intense in the evening, anxious thoughts disrupt ability to sleep. Other psychopathological version of the post-stroke depression is a melancholic depression (24 % cases). Their symptoms are characterized by predominance of effect of melancholy, which is perceived as painful physical suffering, feeling of heaviness, retrosternal pain, sadness and sorrow, which is often followed by inner worry and somatic and vegetative disorders. High intensity of symptoms can be noticed in the morning. Apathetic depressions occur by 24 % of patients. In the clinical picture of the disease deficiency of impulse with decreased vital tonus is predominant. The symptoms depend on diurnal rhythm, which is specific for depression. Their high intensity can be noticed in the morning [2].

Differences of depressions in the psychopathological picture should be considered by choosing of antidepressant treatment. According to the data from literature, tianeptine (known as serotonin reuptake enhancer) is more effective by treatment of melancholic and worried depressions and less effective by treatment of apathetic depressions [2]. Probable causes of the post-stroke depression appear to be [9]:

— patient's reaction on an unexpected disaster, helplessness, restriction of social and professional activity;

— recrudescence of premorbid personality traits (depressive episodes in anamnesis);

— localization of affected area or/and slowing of metabolism in «depressible» zones of the brain (left temporal lobe, thalamus, basal ganglia and limbic cortex);

— imbalance in serotonin and noradrenaline metabolism;

— complication after a long medical treatment (hypotensive, antiarrhythmic, antispasmodic, hormonal and other medicines);

— combination of the above described factors.

It is noticed that by the patients with the post-stroke depression are often discovered changes of a gene, which is responsible for serotonin transporter metabolism. Interaction of genetic and anatomical factors (impairment of definite zones in the brain by the stroke) can play a key role in development of a depressive disorder [10].

In the last 30 years the attention is focused on localization of affected area (impairment of definite zones in the brain by the stroke) [11].

The depression is detected more often by the stroke in temporal lobe and basal ganglia of the left hemisphere, than by impairment of other zones in the left or in the right hemisphere. By impairment of the left hemisphere the early depressive disorder is more likely to develop. In the pathogenesis of the post-stroke depression decrease of monoamines level (serotonin and noradrenaline) is very important. The impairment of serotonergic pathways by the stroke that lead from caudal raphe nucleus and dorsal raphe nucleus to hypothalamus, amygdaloid body, corpus striatum, hippocampus and cerebral cortex, can cause decrease of serotonin level in many parts of the brain. To a certain extent this statement is proved by decrease of level of metabolites of monoamines in cerebrospinal fluid, as also by changes in reactivity of receptors in the left temporal lobe.

The depression is one of the main disorders that can occur after the cerebral stroke. A specific place held by the post-stroke depression in the clinical picture of the stroke first of all is connected with the fact that it is a universal mental response of the patient to different aspects of a stroke development: to the mere fact of the disease, to the organic lesion of the brain, to probable many symptoms of the lesion and to social implications of the stroke. Besides, the post-stroke depression is the main converging point of associations, that appear between this disorder and possible other mental and neurological disturbances, impacting on their clinical implications and finally in large measure determining a short-term and a long-term prognosis of the stroke. For that reason, research of the post-stroke depressions has become one of the major problems in vascular neurology and in neuropsychiatry. It is connected with scientific and practical significance of this issue.

References

1. Capaldi, V., Wynn I. Emerging strategies in the treatment of post-stroke depression and psychiatric distress in patients. Psychol Res Behav Manag 2010;3:109–18.
2. Petrova, E. A., Kontsevoy V. A., Savina M. A.,Nasarov O. S., Skvortsova V. I.// Depressive disorders in patients with cerebral stroke/ Journal of neurology and Psychiatry, 2, 2009
3. Shahparanova, N. V., Kadikov A. S.// Atmosphera. Nervous deseases. 2005. № 3. P. 22.
4. Morris, P. L. P. et al. // Am. J. Psychiatry. 1993. V. 150. P. 124.
5. Downhill, J. E., Robinson R. G. // J. Nevr. Ment. Dis. 1994. V. 182. № 8. P. 425
6. Berg, A., Palomaki H., Lehtihalmes M. et al. Poststroke depression. An 18-month follow up. Stroke 2003; 34: 1: 138–143
7. Vosnesenskaya, T. G.// Hard patient. 2003. B. 1. № 2. P. 26.
8. Vosnesenskaya, T. G. // Selected lectureson neurology/. M., 2006. P. 55–79
9. Kadikov, A. S.// Rehabilitation after stroke / M., 2003.
10. Ramasubbu, R., Tobias R., Buchan A. M. et al. Serotonin transporter gen promoter region polymorphism associated with post-stroke major depression. J Neuropsychiatry Clin Neurosci 2006; 18:96–9.
11. Kanner, A. M. Depression in neurological disorders. Lundbec Inst 2005;161 p.

Agriculture

Innovative Technology of Harvesting of Cotton Stalks

Merdan Shammedov
Turkmen Agricultural University named after S.A. Niyazov, Ashgabat, Turkmenistan

Abstract. *The article is devoted to the problem of the intensification of the measures which are taken in the agricultural industry, requirements for the improvement of the system of cultivation of various kinds of soils and development of new technologies. It results in the economy of facilities, reduction of the labor costs, increase of the fertility of the cultivated lands, as well as in the increase of the cotton production. The intensification of the production processes demands the decision of some problems concerning the perfection of agricultural systems and the creation of new technologies which could promote the increase of the soil fertility and crops production at minimum power and labor costs. The providing of the planned mid-annual production of the cotton-fiber at minimum costs requires a comprehensive mechanization of the cotton growing, and not least the mechanization of harvesting of cotton stalks. The task of harvesting consists in a fast and lost-free gathering of all the mass of cotton stalks as well as in the clearing of fields for the under-winter ploughing.*

Keywords: *cotton, grinder, cotton stalks, technologies, facilities, under-winter ploughing, crop.*

The providing of the planned mid-annual production of the cotton-fiber at minimum costs requires a comprehensive mechanization of the cotton growing, and not least the mechanization of harvesting of cotton stalks. The task of harvesting consists in a fast and lost-free gathering of all the mass of cotton stalks as well as in the clearing of fields for the under-winter ploughing. After the harvesting of raw cotton on fields there are the cotton stalks which should be got in for a short time in order to carry out the ploughing timely. Cotton stalks are used partially as fuel in the country, but the useless majority is burnt, and the remainders are ploughed into the soil. While the machines conduct the uprooting only in part, packing the stalks in rolls, dragging them in heaps and loading in transport carts, the most part of stalks is processed through the plough wholly. Thus the effectiveness and the quality of ploughing decrease. The whole stalks which don't digest during the winter, have a negative influence on the quality of the spring subsoiling, as well as on the harrowing, sowing and intercultivations. In order to avoid it, it is necessary to gather the not digested stalks manually in the spring and to remove them from the field. Besides the ploughing under the infected gummosis, the wilt of stalks provokes the diseases of plants. On the fields where the crop rotations are observed, the chopped during the ploughing cotton stalks in combine with the mineral fertilizer at deep placement promote the yield increase by 4,0 centners per hectare; thus the preparation of

FIG. 1. CCS-3,6 chopper of cotton stalks in field conditions

fields for under-winter ploughing is considerably accelerated.

As a decision of the stated problems a chopper of a simplified construction was designed and tested in the agricultural joint-stock company named after the Hero of Turkmenistan S.Rozmetov (Dashoguz region). During the agricultural work this chopper of cotton stalks CCS-3,6 well proved by tenants and machine operators of the company (figure 1).

The research of the processes of chopping of cotton stalks was began by M.S.Ganiev [1]. The results of his researches were developed further in works by N.A.Kulametov, M.N.Sablikov, R.H.Valeev, I. M.Sablikov and the other scientists [2, 3, 4, 5, 6].

Analyzing the results concerning the considered problem, it is necessary to focus first of all on the works which mention the questions of resource-saving to some extent: power inputs of the process of harvesting and preparation of cotton stalks, decrease of the material consumption and increase of the effectiveness of the gathering.

Due to the fact that the length of cuttings and technology of their chopping have an essential impact on the intensity of the decomposition of cotton stalks in soil, some researches were conducted by us by the chopping of stalks through the working apparatus of chopper CCS-3,6.

The experiment was put in the field. The cotton stalks were selected after the machine chopping and divided into the cuttings of 5, 10 and 15 cm length. The stalks were fractioned at certain humidity and kept within the capron nets. The prepared cuttings of stalks were buried at a depth of 30 cm during the autumn period, and then in 1, 2, 3, 4, 5, 6 months they were dug out, washed and dried. The rate of the decomposition of the organic matter of the stalks embedded into the soil was estimated on the base of the loss of their initial mass from the moment of the placement into the soil. The results of experiments are shown on the Figure 2.

The obtained data testifies that cuttings of stalks in length of 5,0 cm decay the most intensive; the cuttings in length of 10,0 and 15,0 cm show an approximately identical intensity of decomposition.

Based on the above mentioned it is possible to draw the conclusions, that the use of the chopped cotton stalks in combine with the fertilizers is the most economic and energy saving method of the improvement of the organic part of soil.

The cotton stalks in the process of the humification increase the content of humus in the soil and improve its water-physical properties. The cotton stalks contain the nitrogen, phosphorus and potassium twice as much the manure; along with this

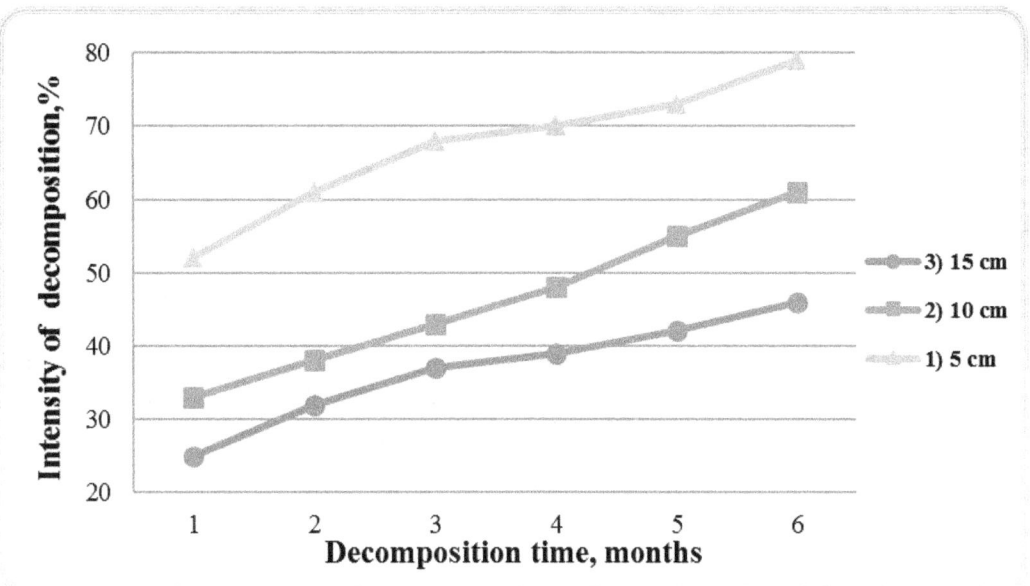

FIG. 2. The intensity of decomposition of stalks depending on the decomposition time in soil
Fractions: 1 - 5.0 cm long; 2 - 10.0 cm long; 3 - 15.0 cm long.

the cotton stalks contain a lot of other macro and micronutrients.

Way back in 1913 R. R. Schroeder got the increase of the yield by 20 percent through the placement of the chopped cotton stalks into the soil, and named the cotton a "happy" plant, which prepares the fertilizer by itself. It should be noted that the entering of the chopped cotton stalks and other plant remains does not replace the fertilizers, although it promotes the creation of the favorable water-physical properties and the optimal conditions for the best growth and development of plants, providing thereby the receipt of high crop yields. The table below shows the contents of basic elements in various plants.

TABLE. The contents of basic elements in various agricultural plants
(in percentage in relation to the air-dry substance: for roots and green mass of raw substance)

Plants, products		Ash elements						
		Nitrogen (N)	Phosphorus (P_2O_5)	Potassium (K_2O)	Magnesium (MgO)	Calcium (CaO)	Sulfur (S)	Total ash
Wheat	grain	2.0-3.0	0.85	0.6-0.9	0.22	0.05	0.18	2.32
	straw	0.6	0.2	0.8-1.0	0.09	0.26	0.05	3.48
Corn	grain	1.8-2.0	0.57	0.37	0.19	0.03	0.06	1.23
	straw	0.75	0.3	1.64	0.26	0.49	0.15	4.37
Rice	grain	1.2	0.81	0.31	0.18	0.07	0.05	5.26
Cotton	seeds	3.00	1.10	1.25	0.54	0.20	0.08	3.90
	fiber	0.34	0.06	0.91	0.17	0.16	0.10	1.93
	bolls	2.54	0.32	3.43	0.28	1.06	0	8.33
	leaves	3.20	0.50	1.28	1.12	6.14	0.40	15.93
	stems	1.46	0.21	1.31	0.41	1.00	0.04	4.54
Beet	sugar	0.24	0.08	0.25	0.05	0.06	0	0.57
	haulm	0.35	0.10	0.50	0.11	0.17	0	1.42

Beet fodder	root	0.19	0.07	0.42	0.04	0.03	0.02	0.86
	haulm	0.30	0.08	0.25	0.14	0.16	0	1.51
Potatoes	tubers	0.32	0.14	0.60	0.06	0.03	0.02	0.97
	haulm	0.30	0.16	0.85	0.21	0.80	0	2.49
Carrot	root	0.18	0.11	0.40	0.05	0.07	0.01	0.93
	haulm	0.34	0.08	0.60	0.15	1.50	0	3.10

References

1. Ganiev, M. S. Technological basis and substantiation of parameters of the working organs of machines for the harvesting of cotton stalks. - Tashkent: the Fan, 1977, - 213p.
2. Kulametov N.A. Technological basis of the mechanization of the harvesting and preparation of cotton stalks. - Tashkent: the Fan, 1990, - 134 p.
3. Kulametov N.A. Development of the technology and the machinery complex for the harvesting and preparation of cotton stalks. The abstract of the dissertation. Moscow, 1991.
4. Sablikov I.M. Substantiation of the parameters of the intensive feeding-chopping apparatus for the stubbing-chopper of cotton stalks. 05. 20. 01 - The abstract of the dissertation. Jangiul, 1995.
5. Sablikov M.N. Choice of the type and definition of the optimum parameters of the working organ for chopping of cotton stalks: The abstract of the dissertation. Tashkent, 1973, - 168 p.
6. Valeev, R. H. Research and substantiation of the technological scheme and parameters of the nourishing device for stubbing-choppers of cotton stalks: The abstract of the dissertation. - Tashkent: 1974, - 17 p.

Other

Academic Background

[1]Anzor Amadaev, [1]Dzhabrail Dasaev, [2]Artur Amadaev
[1]Chechen State University, Grozny, Russia
[2]Russian Academy of science, Grozny, Russia

Network Connection among Web Pages

In the last few years, several works in the literature haveaddressed the problem of dataextraction from Web pages (Baumgartner and Gottlob, 2009b) [1]. The importance of this problem derives from the fact that, once extracted, the data can be handled in a way similar to instances of a traditional database.The approaches proposed in the literature to address the problem of Web data extraction use techniques borrowed from areas such as natural language processing, languages, machine learning, information retrieval, and databases.As a consequence, they present very distinct features and capabilities which makea direct comparison difficult to be done (Summerfield, 2009).With the explosion of the World Wide Web, a wealth of data on many different subjects has become available online.This has opened the opportunity for users to benefit from the available datain many interesting ways. Usually, users retrieve Web data by browsing and keyword searching, which areintuitive forms of accessing data on the Web. However, these search strategies present several limitations (Sahuguet and Azavant, 2008)[2].

Browsing is not suitable for locating particular items of data, because following links is tedious and it is easy to get lost. Keyword searching is sometimes more efficient than browsing, but often returns vast amounts of data, far beyond what the user can handle. As a result, in spite of being publicly and readily available, Web data can hardly be properly queried or manipulated as done, for instance, in traditional databases.For handling Web data more effectively, some researchers have resorted to ideas taken from the database area (Weikum, 2009)[3].Databases, however, require structured dataand, therefore, traditional database techniques cannot be directly applied to Web data. Theadvent of XML as a standard for structuring dataavailable on the Web has brought some light to this problem, but this technology per se does not providea trivial solution for properly manipulating existing Web data. Indeed, the volume of unstructured or semi structured dataavailable on the Web is enormous and is still increasing.Thus, to address this problem, a possible strategy is to extract data from Web sources to populate databases for further handling (Zanasi, 2009)[4].

The traditional approach for extracting data from Web sources is to write specialized programs, called wrappers that identify data of interest and map them to some suitable format as, for instance, XML or relational tables. The most challenging aspect of wrappers is that they must beable to recognize the data of interest among many other uninteresting pieces of text (e.g., mark-up tags, inline code, navigation hints, etc.). These data might havea flat structure, but might also be complex and present an implicit multi-level hierarchical structure that is often non-rigid (Sarawagi, 2008). This means that the data may exhibit structural variations that must be tolerated and treated accordingly (Anton, 2010)[5].

The World Wide Web contains a hugeamount of unstructured data.The need for structured information urged researchers to develop and implement various strategies to accomplish the task of automatically extract data from Web sources (Tanaka and Ishida, 2010) [6].Their ultimate goal is to support business, social or commercial applications. In this paper we presented a survey on the problem of Web automatic dataextraction, focusing on applications, approaches and techniques that were developed during last years.Thereare some future directions and challenges that can be foreseen. Some of them comprise how to address enormous scaling issues of theextraction problem, the robustness of the process and the design and implementation of auto-adaptive wrappers (Winograd, 2010)[7].

Related Work

While leading this research the number of interesting works is encountered. The work which is attracted my attention was Schickler's, Mazer's and Brooks's which is called "Pan-Browser support for annotations and other meta-information on the World Wide Web." This work emphasizes advanced approach for group of people to establish and share information about the documents content which can be reached via the World Wide Web. The main interest in this research is supporting asynchronous communication between users to provide mechanism for group discussions, where every single user will be able comment on any documents on particular Web server. Roscheisen and Winograd (1995) suggested that user-created meta-information might not be used only in ways of annotations, but also in voting, trails which support notions in a frequently uninspected communication average. This system does not depend on changes in Web browsers and servers. The main functionality of the system is given by a specialized proxy (Qiao and Yoo 2000)[8]. Furthermore, the commentaries can be written not only at the end of the documents, but in any places where the author wishes. In contrast to other previous approaches this mechanism of creation, visualising and control of commentaries written by users might be used without identifying any browser or server.

The Computer Science scientific literature counts many valid surveys on the Web dataextraction problem: Laender et al. (Laender et al. 2002)' in 2002, presented a notable survey, offering a rigorous taxonomy to classify Web dataextraction systems.They introduced a set of criteriaand a qualitativeanalysis of various Web dataextraction tools.

In the same year Kushmerick (2002) tracked a profile of finite-stateapproaches to the problem, including analyzing wrapper induction and maintenance, natural language processing and hidden Markov models. Kuhlins and Tredwell (Kuhlins and Tredwell 2003) surveyed tools for generating wrappers already in 2003: information could not be up-to-date but analyzing theapproach is still very interesting. Again on the wrapper induction problem, Flescaet al.(Flescaet al. 2004) and Kaiser and Miksch (Kaiser and Miksch 2005) surveyed approaches, techniques and tools.The latter in particular modeled a representation of an Information Extraction system architecture.

Chang et al.(Chang et al. 2006) introduced a tri-dimensional categorization of Web dataextraction systems, based on task difficulties, techniques used and degree of automation. Fiumara (Fiumara 2007) applied these criteria to classify four new tools that arealso presented here. Sarawagi published an illuminating work on Information Extraction (Sarawagi 2008): anybody who intends to approach this discipline should read it. To the best of our knowledge, the newest work from Baumgartner et al. (Baumgartner et al. 2009)[9] is a short survey on the state-of-theart of the discipline.

Network Structure

According to Newman and Barabási (2005) most web portals consist of a stable skeleton, which represents the overall organisation of the web portal. The number of documents only temporally linked to the skeleton. Each news item stands for individual web document with a specific URL. Then this item is added to the main page and to the web page categories which it belongs to. For instance, the release of the new movie could begin on the front page, the movies page, and the category of movie subdirectory of the movies page. Plake (2006) says that the documents which belong to the skeleton qualified by daily visitation model. Therefore, the number of visitors who access the web document rises in specific time. It is also important to find the difference between two visitation models, it allows to recognise the web pages, which belong to the skeleton from the news documents (Newman and Barabási , 2005). To implement this, the researches made accumulative visitation model and computed the deviation from the suitable lines. Then the documents were randomly selected and checked, and the results showed that news document and skeleton mostly have the different format. The skeleton of analysed webpage by researches (Albert and Barabási, 2001) had 933 documents, network had a couple of strongly – linked hubs, while many others were connected to a skeleton by an individual link.

Network Visitation

An interesting work about dynamics of network visitation is done by scientists of University of Notre Damme, USA. Knowing that the difference between news documents and skeleton is forced by models of visitation, the research is mainly based on interaction between single users and the general visitation of a document. The general visitation of a particular document is defined by the document's position on the website and the contents likely significance for different user groups. Nevertheless, there is some decay in visitation of web page which is caused by users who have already visited the page and read the document which they are interested in, will not do it again. Therefore we can see some decrease of visits to particular website with time. But, if the content of webpage will be upgraded and modified, there is possibility of increasing the visits of users. The visitation of a document can also be determined by the number of new users visiting the website where the document is released (Dorogovtsev and Mendes , 2003). Furthermore, it is also important to fix display time of a given document on web page, which shows amount of users who did not manage to read this document by not visiting the web site when the document was displayed. Here is some graphical representation of the studied web portal (Fig. 1).

Even though we think that the visitation of the particular document on web page depends on its popularity, the results show that the dynamics of its effective accessibility also play an important role. These results are not just suitable for news portals, but also for commercial web sites. These kinds of approaches to define the rank of visitation to web pages not only give us the understanding of

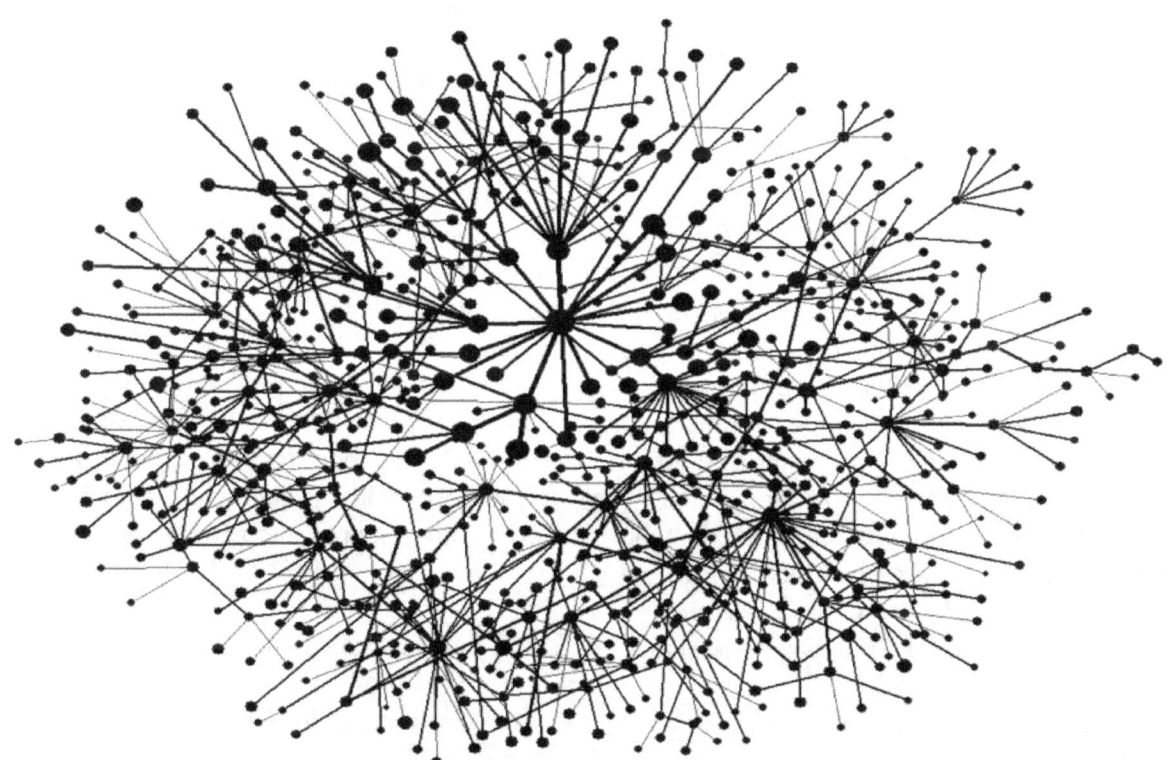

FIG.1. The central largest node is the main page which is connected to other most frequently visited pages. Width of the links mean that this hyperlink was used mostly to visit the specific page.

information access, but also important ideas about the web page design to make it more attractive to users.

Web Data Extraction Systems

We can generically definea Web dataextraction system as a sequence of procedures that extracts information from Web sources (Laender et al. 2002). From this generic definition, we can infer two fundamental aspects of the problem:
-Interaction with Web pages
-Generation of a wrapper

Baumgartner and Flesca (2009) definea Web dataextraction system, as 'a softwareextracting, automatically and repeatedly, data from Web pages with changing contents, and that delivers extracted data to a database or some other application'.

This is the definition that better fits the modern view of the problem of the Web dataextraction as it introduces threeimportant aspects:
- Automation and scheduling
-Data transformation, and the
-Use of theextracted data

The following five points cover techniques used to solve the problem of Web dataextraction.

Interaction with Web pages

The first phase of a generic Web data extraction system is the Web interaction (Wang et al. 2011): Web sources, usually represented as Web pages, but also as RSS/ Atom feeds (Hammersley 2005)' Microformats (Khareand Qelik 2006) and so on, could be surfed by users, both in visual and textual mode, or just simply inputted to the system by the URL of the document (s) containing the information.

Some commercial systems, Lixto for first but also Kapow Mashup Server (described below), includea Graphical User Interface for fully visual and interactive navigation of HTML pages, integrated with dataextraction tools.

The state-of-the-art is represented by systems that support theextraction of data from pages reached by deep Web navigation (Baumgartner et al. 2009)' i.e. simulating theactivity of users clicking on DOM elements of pages, through macros or, more simply, filling HTML forms.

These systems also support theextraction of information from dynamically generated Web pages, usually built at run-timeas a consequence of the user request, filling a template page with data from some database. The other kind of pages are commonly called static Web pages, because of their static content.

Generation of a Wrapper

Just for now, we generically define the concept of wrapper as a procedureextracting unstructured information from a sourceand transforming them into structured data (Zhao 2010; Irmak and Suel 2006). A Web dataextraction system must implement the support for wrapper generation and wrapper execution. We will cover approaches and techniques, used by several systems, later.

Automation and Scheduling

The automation of page access, localization and extraction is one of the most important features included in last Web data extraction systems (Phan and Ho 2005): the capability to create macros to execute multiple instances of the same task, including the possibility to simulate the click stream of the user, filling forms and selecting menus and buttons, the support for AJAX technology (Garrett 2005) to handle the asynchronous updating of the page, etc. are only some of the most important automation features.

Also the scheduling is important, e.g. if an user wants to extract data from a news website updated every 5 minutes, many of the last tools let him to setup a scheduler, working likea croti, launching macros and executing scripts automatically and periodically.

Data Transformation

Information could be wrapped from multiple sources, which means using different wrappers and also, probably, obtaining different structures of extracted data. The steps between extraction and delivering are called data transformation: during these phases, such as data cleaning (Rahm and Do 2000) and conflict resolution (Monge 2000)' users reach the target to obtain homogeneous information under a unique resulting structure.

Most powerful Web dataextraction systems provide tools to perform automatic schema matching from multiple wrappers (Rahm and Bernstein 2001)' then packaging datainto a desired format (e.g. a database, XML, etc.) to makeit possible to query data, normalize structureand de-duplicate tuples.

Use of Extracted Data

When theextraction task is complete, and acquired dataare packaged in the needed format, theseinformation are ready to be used; the last step is to deliver the package, now represented by structured data, to a managing system (e.g. a native XML DBMS, a RDBMS, a data warehouse, a CMS, etc.). In addition to all the specific fields of application covered later in this work, acquired data can bealso generically used for analytical or statistical purposes (Berthold and Hand 2008) or simply to republish them under a structured format.

Applications

On the one hand the Web is moving to semantics and enabling machine-to-machine communication: it is a slow, long term evolution, but it has started, in fact. Extracting data from Web sources is one of the most important steps of this process, becauseis the key to build a solid level of reliable semantic information. On the other hand, Web 2.0 extends the way humans consume the Web with social networks, rich client technologies, and the consumer as producer philosophy. Hence, new evolvements put further requirements on Web dataextraction rules, including to understand the logic of Web applications.

In the literature of the Web dataextraction discipline, many works cover approaches and techniques adopted to solve some particular problems related to a single or, sometimes, a couple of fields of application. Theaim of this section, at the best of our knowledge, for the first time, is to survey and analyze the greatest possible number of applications that are strictly interconnected with Web dataextraction tasks. In the following, we describea taxonomy in which key application fields, heavily involved with dataextraction from Web sources, are divided into two families, enterpriseapplications and social applications.

Summary

The World Wide Web contains a huge amount of unstructured data.The need for structured information urged researchers to develop and implement various strategies to accomplish the task of automatically extract data from Web sources. Their ultimate goal is to support business, social or commercial applications. In this section we presented a survey on the problem of Web automatic data extraction, focusing on applications, approaches and techniques that were developed during last year's.There are some future directions and challenges that can be foreseen.Some of them comprise how to address enormous scaling issues of the extraction problem, the robustness of the process and the design and implementation of auto-adaptive wrappers.

Web Development Tool Used

Python programming language is used as a web development tools to extract data form web pages. Python is a dynamically typed language with a rich set of native types.Its number hierarchy includes native arbitrary-length integers, hardware-precision floating-point and complex numbers, and library support for rational numbers and arbitrary precision floating point. It also has powerful strings, variable-size lists, sets, and very flexible associative arrays called dictionaries in Python.These types give Python a rich vocabulary in which to express many complex algorithmic questions with clarity and efficiency.We can illustrate this expressive power with a fully functional implementation of the classic QuickSort. Python combines high-level flexibility, readability, and a well-defined interface with low-level capabilities, including an official C interface that lets you extend the language with C code and link to third-party libraries in C, C++, and Fortran (Zhai and Liu, 2010).

This generality is advantageous for modern scientific computing: it's a productive everyday environment that also lets you optimize performance-critical bottlenecks. Further, it provides the flexibility to build tools with a precise balance of low- and high-level features so you can appropriately choose between performance and ease of development or use. This combination of semantic richness and flexibility makes Python well suited to solving many of the non-algorithmic computational issues we mentioned above, such as integrating with the Web, data formats, or low-level hardware (Zhao, 2010). Python libraries—whether included in the language or from external projects—let you interface with Web servers, databases, and scientific storage formats such as HDF5 to process text and send data over raw network sockets, and so on. Such tasks often require a mix of fairly low-level machinery and high-level abstractions, such as the combination of managing binary data over raw sockets with objects to represent Web servers or open database connections (Pilgrim and Mark, 2009).

Web Connectivity Analysis

The links connecting documents in the web arein principleall equivalent: the web itself does not express an preference for one link or one document aboveanother. Yet, the connectivity or pattern of linkages between pages does contain a lot of implicit information about the relative importance of links. The author of a web document will normally only include links to other documents that are relevant to the general subject of the page, and of sufficient quality. Thus, locating one document relevant to your goals may be sufficient to guide you to further information on that issue. High quality documents, that contain clear, accurateand useful information, are likely to have many links pointing to them, while low quality documents will get few or no links. Thus, although no explicit preference function is attached to a link, thereis a preferenceimplicit in the total number of links pointing to a document.This preferenceis produced collectively, by the group of all web authors.

Thereexist different mathematical techniques to extract this information. Recently, two types of algorithms have been developed for this purpose: PageRank (Bossa and Provetti 2006) and HITS (Kaiser and Miksch 2005). Both usea bootstrapping approach: they determine the quality or "authority" of a web page on the basis of the number and quality of the pages that link to it. Since the definition is recursive (a page has high quality if many high quality pages point to it), the algorithm needs several iterations to determine the overall quality of a page. Mathematically, this is equivalent to computing the eigenvectors of the matrix that represents the linking pattern in the selected part of the web. Page Rank uses the linking matrix directly, HITS uses a product of the matrix and its transposed matrix. The latter method produces two types of pages: authorities, that are pointed to by many good "hubs" (indexes or lists of web pages), and hubs, that point to many good authorities. In combination with a keyword search, which restricts the pages for which the quality is computed to a specific problem "neighborhood", these methods seem to producea much better quality in theanswers returned for a query.

The disadvantage of these methods is that they are static: they merely use the (rather sparse) linking pattern that already exists; they do not allow the web to adapt to the way it is used, as the learning web algorithms propose. However, the two methods can complement each other, as the use of connectivity matrices does not require these matrices to have only binary values (either thereis a link or thereis not).The learning web and other techniques will produce less sparse matrices with numerical values that can be analysed in the same way, but are likely to produce more fine-grained and reliable results.

References

1. Baumgartner, R., Campi, A., Gottlob, G., AND Herzog, M. (2010), "Web data extraction for service creation", Search Computing: Challenges and Directions.
2. Baumgartner, R., Ceresna, M., and Ledermuller, G. (2009), "Deepweb navigation in web data extraction", In CIMCA '05: Proc. of the International Conference on Computational Intel- ligence for Modelling, Control and Automation and International Conference on Intelligent Agents, Web Technologies and Internet Commerce Vol-2 (CIMCA-IAWTIC'06). IEEE Computer Society, Washington, DC, USA, 698-703
3. Baumgartner, R., Flesca, S., and Gottlob, G. (2009a). "The elog web extraction language". In LPAR '01: Proc. of the Articial Intelligence on Logic for Programming. Springer-Verlag, London, UK, 548-560
4. Baumgartner, R., Flesca, S., and Gottlob, G. (2009b), "Visual web information extraction with lixto", In VLDB

'01: Proc. of the 27th International Conference on Very Large Data Bases. Morgan Kaufmann Publishers Inc., San Francisco, CA, USA, 119-128

5. Baumgartner, R., Froschl, K., Hronsky, M., Pottler, M., And Walchhofer, N. (2010).

6. Baumgartner, R., Gatterbauer, W., And Gottlob, G. (2009). "Web data extraction system", Encyclopedia of Database Systems, 3465-3471

7. Baumgartner, R., Gottlob, G., And Herzog, M. (2009). Scalable web data extraction for online market intelligence. Proc, VLDB Endow. 2, 2, 1512-15 23.

8. Bossa, S., Fiumara, G., And Provetti, A. 2006. A lightweight architecture for rss polling of arbitrary web sources. In WOA.

9. Berthold, M. And Hand, D. J. (2008). Intelligent Data Analysis: An Introduction.

Forecast of potential natural risks for the historical and architectural sights of the Holy Mountains Lavra (Sviatohirsk Monastery)

Valeriy Sukhov

Kharkiv National University named after V.N. Karazin, Kharkiv, Ukraine

Abstract. *The article is devoted to the forecasting of the potential risks to the monuments of nature, history and architecture of the Sviatohirsk monastery. It considers a comprehensive approach to the identification of the key factors of the natural hazards for the old buildings and constructions of the monastery which are built on the loamy and chalky carbonate rocks.*

Keywords: *natural risks, suffusion, karst, seismicity, carbonate rocks, groundwater, Sviatohirsk monastery.*

Introduction

In the world practice the risk is considered to be a measure of danger to the existence and inviolability of any object, among which the most important are the monuments of history and architecture. It is the probability of threats in the form of some natural or man-made phenomena, which manifestations lead in particular to the destruction of buildings and structures. Risks of the natural hazards which are associated with the geological and especially geodynamic processes and phenomena (such as suffusion, karst, landslides, earthquakes etc.) are difficult to forecast in most cases, and for this reason they are considered to be dangerous. Most of them are impossible without the participation of natural waters as one of the two main components of the hydrolytic Earth crust system [1].

A well-timed assessment of the hazards which are associated with the geological factors is of great necessity in order to prevent the possible negative impact of the natural geological processes on the different objects, including the cultural and business ones. The integrated approach which is intended to the solution of this problem can include four main stages [2]:

1. System analysis and identification of the key issues of the natural hazards.

2. Definition of the possible and actual impact of the individual and complex geodynamic processes on the protected objects.

3. Calculation of the degree of danger in the event of the emergencies which are related to the changes of the geological parameters.

4. Development of measures which are aimed at minimizing of the negative impact of the geological

(geodynamic) processes.

As far as the assessment of any natural geological risk in each case cannot always rely on the sufficient amount of factual material, some of our conclusions and recommendations are approximate or probabilistic in nature.

Let's consider the problem on the example of Sviatohirsk Monastery (Donbass, Ukraine), which was built in 16-17th centuries on the loamy-chalky rocks of the Upper Cretaceous. The monastery buildings are located on the chalk cliffs up to 100 meters high on the right bank of the Seversky Donets River, which passes through the channel zone of the tectonic active Petrovsk-Kremennaya fault.

Main part

The geodynamic processes which occur within the Sviatohirsk monastery can be divided into the local and regional ones. The most devastating among the local processes for the marly and chalky strata are the suffusion and karst.

The significant natural geological risks among the regional processes of the researched territory are associated with the earthquakes of magnitude 4 on the Richter scale; during the 20th century the seismic centers of these earthquakes were recorded at a distance of 40-50 km from the monastery. Since the monastery was built without any consideration of the seismic threat, in the case of this geodynamic phenomenon repetition near the monastery, it can be devastating for the buildings of the complex. This becomes even more important with the ancient constructions, which are located on the top of the "chalky cliff", particularly with St. Nicholas Church.

As to the geodynamic processes, the development of the suffusion areas on the researched territory are defined as the places of the surface discharge and the intensive underground runoff of the infiltration waters. The latter problem is mainly a turbulent movement of the water through the open cracks, which occurs both in saturation mode (during the periods of the high water content), and in unsaturated mode of free filtration. Sometimes there is the water movement in channels on subvertical open cracks, which are more than 10 cm. In such cases the infiltration moving water is in the form of the gravitational free fall, which causes the largest underground mechanical destruction of the carbonate rocks [3].

Concerning the array of the "Cretaceous rocks" the author distinguishes two main types of the suffusion – the surface one and the underground one. When the first one is associated with the atmospheric precipitation, i.e. with the surface waters, the second one is associated with the infiltrated waters within the fracture and pore space of the loamy and chalky strata and, therefore, is groundwater structurally.

In addition to the width and depth of the exogenous cracks, the Quaternary alluvial-delluvial deposits (soils) are considered to be an important factor of the groundwater suffusion in carbonate thickness of the region. Meeting these soils on the way the infiltration flows are assimilated with the loose Quaternary sediments, which are represented here by yellow-brown loams, sands, subsands, chalk, gravel and humus, almost completely. Reaching 70-80% of water saturation, these "cover" deposits inhibit infiltration of the traffic flows, transferring them then from the surface to the groundwater.

The business activities which causes the stream erosion, landslides, avalanches and small debris in the areas of the trimmed slopes and destruction of the vegetation has also some impact on the development of the geodynamic processes in Sviatohirsk monastery.

During the periods of heavy rains and snow break the trodden paths serve often as the temporary flows that deplete the slopes rapidly. Thus, the devastating effects of the stream erosion, which is a kind of the surface suffusion, are clearly observable near St. Nicholas Church and on the eastern side of the "chalk cliffs". The landslides caused by the mechanical ground water flow can cause the destruction of the buildings of the monastery complex.

In the rear part of the old monastery walls there is a suffusion crater which was formed as a result of the intensive groundwater circulation in the system of the interlayer space "soils and chalky rocks"; the rate of the filtration is there the highest one because of the presence of coarse debris weathering products. The monastery walls are permanently destroyed because of the disrepair of the old drainage network. In its turn, the destruction products cover the stacking that ten times increases the intensity of the suffusion processes and leads to the cracking, bulging of walls and the collapse of the old building structures.

The patterns of distribution of the local geodynamic risks and, in the main, the suffusion and karst on the territory of Sviatohirsk monastery are closely related to the areas of the infiltration and movement of groundwater within the array of the loamy and chalky rocks and are determined by the dynamics of the latter. All the above-mentioned, including the influence of both geological and anthropogenic factors, allowed the identifying of the main areas where the local geodynamic processes are the most dangerous for the monuments of history and architecture. Among such sites within the "chalk cliffs" are the following:

- The summit on which St. Nicholas Church stands (suffusion, epikarst, hypokarst);
- The eastern part (suffusion, epikarst);
- The lower part (suffusion, epikarst, hypokarst, landslides, river abrasion).

The seismic activity in the region constitutes the overall danger to the researched object.

Among the geological risks which may threaten the buildings and structures of the historical and architectural complex of the Sviatohirsk monastery seriously, it is advisable to identify the main ones which are related to the karst-suffusion and the present-day seismic processes. They are based on the process of the continuous transformation of the hydrolithspheric matter in solid, liquid and gaseous envelopes of the planet. They lead not only to the changes in the status of the individual components of the Earth crust but also to the interaction of the geophysical and geochemical fields of the different nature. These processes underlie the global geodynamics of the Earth and the development of endogenous, exogenous, hydrogeodynamic, hydrogeochemical and other phenomena which have a significant impact on the geotechnical parameters of individual sections [2].

The calculation of natural risks includes the social, physical and economic losses. The prediction of such losses on the one hand allows the forecasting of the negative consequences of the geodynamic processes, on the other hand it allows the developing of a rational set of measures which are intended to the minimization of their impacts.

The suffusion and karst are the geodynamic processes caused by the mechanical and chemical destruction of the carbonate rocks, foundations, and in some case of the walls, respectively (St. Nicholas Church and the underground structures of "Chalky rocks").

The karst and suffusion dangers are associated with the deformities of the carbonate rocks; in the world practice it is taken as the average perennial intensity of high intensity carst formation [4]. In the majority of cases the regional assessment of the suffusion and karst hazard is qualitative. It is based on the analysis of the main factors of the development of suffusion and karst as well as on the special zoning affect by the degree and intensity of these processes [2].

There are different methods of calculation of the karst and suffusion risks [5, 6]. And every of them are of the probabilistic nature. Therefore, it is advisable to determine the potential danger of these processes through the easiest method with the help of the following formula [3]:

$$P = \frac{(S_{ck} - S_g)}{S}, \quad (1)$$

where P is the index of the potential danger (risk) of karst and suffusion; S_{ck} is the area of the karst and suffusion processes which could be caused by the deformations of the Earth surface on the territory of a researched km²; S_g is the area of the existing karst and suffusion deformations on the territory of a researched km²; S is the area of the research in km².

The value of the indicator of risk (RR (s)) may vary from 0 to 1.

For the main research components of the formula mean values are equal: S_u is 0,6 km²; S_c is 0,61 km²; S is 1 km², where

$$P_p(c) = \frac{(0,6-0,1)}{1} = 0,5. \quad (2)$$

These figures mean that the rate of the karst and suffusion potential hazard to the monuments of history and architecture of the Sviatohirsk monastery is quite high and is 0.5. According to the modern risk theory, this figure is the expectation of the specific physical risk of the karst and suffusion territory deformations [4]. It reproduces not only the level of the natural hazards mathematically, but also creates a basis for the evaluation of the possible social and economic damage to the buildings of the monastery complex.

The earthquakes are the tremors and vibrations of the Earth surface which are caused by the sudden release of the potential energy of the Earth interior, and which are characterized by the elastic vibrations which propagate in the lithosphere, i.e. the seismic waves. The depth of the earthquake foci, which are divided on the base of the cause into the tectonic, volcanic, dam and sea ones, vary from 1 to 700 km. On the platform part of Ukraine, to which the researched area belongs, there are seismically active zones; one of them covers the graben of the Western Donetsk. The seismic activity reaches here 4-4.5 points on the Richter scale and is classified as a moderate one. The cells of earthquakes according to the seismic studies [7] are located at shallow depths, i.e. up to 10 km, and are possibly related to the tectonic movements in the geologically ancient closed transverse faults which cross the Dnieper-Donetsk paleorift.

Such earthquakes which are registered to the north and to the east of the Sviatohirsk Monastery, do not cause a major damage, but can cause a serious damage to the monuments of history and architecture which are located on its territory. Mainly it would hurt the chalky St. Nicholas Church and underground places of worship, as far as even minor by force earthquakes cause landslides and avalanches [8].

Today there are a lot of ways of the assessment of the earthquakes risk, from the building of three-dimensional models of the seismic influence to the mathematical calculations of the varying complexity. In order to assess the seismic risks on the researched territory the author used the simplest one. The seismic effects on the buildings of the Sviatohirsk monastery are determined through the empirical relationship, known as the formula of the macroseismic field or the formula of the seismic impact intensity [2], used by the author in simplified form, i.e. without taking into account the regional factors-:

$$I = M - \lg\sqrt{A^2 + h^2}, \qquad (3)$$

where I is the impact of the seismic intensity of the earthquake; M is the magnitude of the earthquake; A is the distance of the focus from the researched object; h is the depth of the focus of the earthquake.

If the magnitude (M) of the possible earthquake on the researched area is 4, the depth of the focus is about 10 km and the average distance of the focus from the Sviatohirsk Monastery is 40 km, the intensity of the seismic effects, determined by the possible destruction will be:

$$I = 4 - \lg\sqrt{40^2 + 10^2} = 2,38. \qquad (4)$$

According to the mathematical value of the intensity of the seismic influence we can determine the likely extent of damage to the building structures. All the oldest buildings on the territory of the monastery were built without regard to the antiseismic measures, with torn or chalky stone on lime or cement grout. Some constructions (such as walls, stairs, and underground galleries) were built with bricks on lime or cement. According to the classification of risk types it conforms to the level A1 and A2 [4]. The modified through the international seismic scale MMSK-86 seismic effect on the researched object is 2.38. In case of the repeated earthquakes with the same focus, the buildings and structures of the Sviatohirsk monastery may suffer a moderate damage, namely the damage to material and non-constructive elements of the buildings, such as the falling away of the plaster layers, cracks in walls, the damage to the main weak structures in the form of cracks in the walls, as well as the deformation at the junction panels. In order to eliminate the damage it will be necessary to have an expensive repair of the architectural forms [2].

Thus, as the suffusion and karst processes occur constantly in "chalky rocks", the possible earthquakes are significant indicators of the natural geological risks which threaten the integrity of the buildings of the Sviatohirsk monastery complex.

Conclusions

1. The geodynamic processes within the researched territory are divided into the local and the regional ones, which are caused by seismic,

tectonic, structural, hydrogeological, lithological, geochemical and other geological features.

2. Among the local geodynamic processes the suffusion and karst which are formed in the natural system of "rock water" are the most widely observed, and among the regional ones this is the seismic activity of the researched territory which is associated with the tectonic activity of the deep faults.

3. The estimated potential geological risks suggest that the indicators of the potential danger from the karst and suffusion (0.5) and seismic impact (2.38) are quite high for a possible damage to the materials and buildings of the Sviatohirsk monastery.

References

1. Knight F. Risk. Uncertainty and Profit. Boston: Haughton Miffin Co., 1921. – P. 210–235.
2. Lysychenko G.V., Zabulonov Y.L., Khmil G.A. Natural, technological and environmental risks: analysis, evaluation, management. Kiyv: Naukova Dumka, 2008. - 541 p.
3. Klimchuk A.B. Epykarst: hydrogeology, morphogenesis and evolution. Simferopol: Sonata, 2009. - 112 p.
4. Rohozyn A.L., Elkin V.A. The regional evaluation of hazards and risk caused through the karst. Problems of the security and the emergency situations, 2003, № 4. - P. 33-52.
5. Khil G. Assessment of the potential man-made and natural security of the territories of Ukraine based on the system analysis. Environment and Resources: Collection of scientific works. Institute national security issues. Kiyv: IPNB, 2007. Vol. 17. - P. 54-65.
6. Akimov V.A., Radaev N.N., Sakharov M.V. Determination of territories of the relative hazards. Problems of security in the emergency situations, 2000, № 6. - P. 129-140.
7. National Atlas of Ukraine. Kiyv: DNVTS "Mapping", 2009. - 440 p.
8. Rud'ko G.I., Pavliv N.P. Risk of the development of some dangerous geological processes in the Carpathian region of Ukraine. Visnyk of L'viv University. Series: Geology, 2001. Vol. 15. - P. 49-53.

www.ingramcontent.com/pod-product-compliance
Lightning Source LLC
Chambersburg PA
CBHW080919170526

45158CB00008B/2166